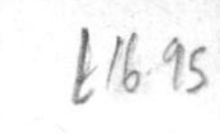
£16.95

D0235547

MECONOPSIS

Christopher Helm
HARDY PLANT Series

ORNAMENTAL GRASSES *Roger Grounds*

Forthcoming:

CAMPANULAS *Peter Lewis and Margaret Lynch*

HARDY EUPHORBIAS *Roger Turner*

ALLIUMS *Dilys Davies*

DIANTHUS *Richard Bird*

James L. S. Cobb

Published in association with the Hardy Plant Society

CHRISTOPHER HELM

London

TIMBER PRESS

Portland, Oregon

Line illustrations by Andrew Hutchinson

Christopher Helm (Publishers) Ltd, Imperial House,
21–25 North Street, Bromley, Kent BR1 1SD

ISBN 0-7470-0427-7

A CIP catalogue record for this book
is available from the British Library

First published in North America
in 1989 by
Timber Press, Inc.
9999 S. W. Wilshire
Portland, Oregon 97225

ISBN 0-88192-151-3

Library of Congress Cataloging-in-Publication Data

Cobb, James L. S.
The genus Meconopsis / by James L. S. Cobb.
p. cm.
Bibliography: p.
Includes index.
ISBN 0-88192-151-3
1. Meconopsis. I. Title.
SB413.M42C6 1989
635.9'33122—dc20 85-5009
CIP

Typeset by Florencetype Ltd, Kewstoke, Avon

Printed and bound in Great Britain by Biddles Ltd, Guildford and Kings Lynn

For Calla, the perfect gardener's companion

Contents

Colour Plates

Figures

Acknowledgements

The author would like to express his gratitude for the time and effort the following people devoted to advising on various aspects of this book. Mike Hirst of Durham Agricultural College, Mike and Polly Stone of Fort Augustus, Dr Mary Noble, Margaret and Henry Taylor of Invergowrie, David Tattersfield of Branklyn Gardens, Brenda Anderson of Balruddery, Mr and Mrs G.A. Williams of Vancouver, Francis Cabot of Cold Spring, New York, and Bob Mitchell, Sandy Edwards and Dr Ruth Ingram of St Andrews University. The author would also like to record his indebtedness to Henry Taylor (Plate 5), Ron McBeath (Plates 3 and 6) and Peter Cox (Plate 10), and to Andrew Hutchinson for the delightful line drawings of flower features.

Preface

The Himalayan poppies, properly known as the genus *Meconopsis*, are a group of plants that have flowers up to 28 cm (11 in) across and show in perfect purity all the colours imaginable. They flower from spring to early autumn and contain species that would grace the smallest rock garden as well as the most overgrown woodland glade. The winter foliage can have all the diversity of shape and form as do snowflakes. One would imagine a group of plants with these talents would be very popular. The fact that they are not widely grown requires some explanation, and, to encourage a change of attitude, the removal of some prejudices. It is not that they are unduly difficult, after all they have been used as bedding plants in Hyde Park in London and at Ibrox in Glasgow. The reason in part is probably related, not to the innate difficulties of the plants, but to the failing of people to understand the basic needs of the easier sorts.

Reginald Farrer, the father of the rock garden, certainly put his finger on a major problem—the Latin names. *Meconopsis* itself is a dinosaurish type of name likely to be associated with the stuffier sort of herbarium. Farrer tried his best to sort the problem out by giving different species such names as the 'lonely poppy', the 'celestial poppy' or 'Farrer's lampshade poppy'. He became typically quite worked up over *M. quintuplinervia*; regrettably no one has taken much notice of Farrer and we still shuffle around the garden pointing to a 'good form of quintuplinervia'. Calling them blue poppies does not help much either since it rather smacks of such impossibilities as blue carnations created with coloured dyes. Meconopsis were particularly associated with gardeners of the 1930s who used to appear at the Chelsea Flower Show and the like with newly discovered species which only they were expert enough to grow. The mystique has grown up that they can only be cultivated in a few Scottish gardens. This needs dispelling!

Many of the most famous Scottish gardens are on the east coast, including the most famous of them all, the Royal Botanic Garden in Edinburgh. Rainfall is not high there; it is the wrong side of the country. It can be bitterly cold without snow cover, it is very rarely humid in summer and periods of real drought can be extensive and unpredictable. Sunshine is more frequent than on the west but one has to admit that it is rarely subject to temperatures as high as 27°C (80°F). Sweet cherries and apricots, however, grow outside in my garden just north of Edinburgh and my meconopsis are shaded by half-a-dozen established species of *Eucalyptus*. There is nothing magical about the climate in Scotland, it merely is a useful baseline to showing how to grow meconopsis.

The genus *Meconopsis* is almost synonymous with Sir George Taylor who still lives in Dunbar in direct line of sight across the Firth of Forth from where I live in Fife. He worked on this genus at a time when a great deal of new plant material was coming in from the Himalayas and China, and eventually produced in 1934 what to this day is still one of the finest monographs on a single genus (Taylor 1934). It set a standard that has rarely been equalled, though it must be confessed that the lengthy and quite excellent section on cultivation was written by E.H.M. Cox who had masterly practical experience in growing them. Cox lived at Glendoick in Perthshire, where his son Peter now runs the world-famous rhododendron nursery and who himself has recently been plant collecting in China. Taylor revised the earlier classification of Prain and in doing so showed a rare mixture of the expertise of the plantsman and the taxonomist; his classification has stood the test of time. Taylor later joined one of the Frank Ludlow and George Sherriff expeditions to Bhutan and, although he was ill, the 16 species he found in the field confirmed his taxonomic ideas. Taylor subsequently wrote of his findings and these little-read articles are extremely illuminating. His later studies led him to remove two species from his original classification. *M. florindae* and *M. georgiae* were classified on the strength of yellow flower colour separating them from the similar blue-flowered species. Ludlow and Sherriff and later Taylor himself then found a yellow form of the widespread *M. horridula* and, although this is a little difficult to explain in terms of genetics and evolution, Taylor realised that flower colour was not a sufficient criterion for definition of a species.

Taylor recognised 41 species and subsequently removed the two mentioned above but added pink *M. sherriffii*. Since that time Williams added *M. taylorii* from Nepal and the Chinese have added at least another four species. This makes a total of 45 known species. It must be likely that in the remote areas of south-east Tibet and western China itself there are others since many species of meconopsis have very limited distribution. There are certainly likely to be many forms, with desirable characteristics, of species that are already described. The distribution of *Meconopsis* is shown in Figure 3, p.8.

The Chinese are particularly interested in plants with medicinal or agricultural potential and although the seeds of some species have been pressed for oil and the buds and young shoots eaten as greens, meconopsis have not been exploited. The main threat from a conservation point of view is pressure of grazing but as they have coped with this for many centuries it would not seem an outstanding problem. All species, even the most perennial, renew themselves from abundant seed and would probably rapidly re-establish in areas temporarily taken into cultivation or damaged by fire.

I am not quite sure why I became involved in meconopsis cultivation, except perhaps because of the words of an honoured academic in my student days who advised that success was often more certain in science if one chose to be a world authority on something obscure. My first encounter with meconopsis, other than the wildling *M. cambrica* which seeded itself in the wild corners of my earliest childhood garden, was a plant of *M. betonicifolia* that came with a house my parents moved into in my late teens when I was already well hooked on horticulture. This plant was I think considered to be a major selling point. In retrospect it was a pretty poor specimen living a misunderstood life under a fir tree.

Difficult plants of course attract a certain type of gardener and my rockery is full of things that are in terminal decline, and half my apple trees are unusual varieties that rarely fruit and never ripen. It has to be admitted that the sight of drifts of *M.grandis* underneath the trees in the Edinburgh Royal Botanic Garden would fire enthusiasm in anyone with the slightest horticultural stirrings, and the whole aim of this book is to convince you that *Meconopsis* is not particularly difficult to grow.

Introduction

The genus *Meconopsis* consists of poppies and thus the various species are members of the family Papaveraceae. A flower of the native British wildling *Meconopsis cambrica* and, say, a flower of the Iceland poppy *Papver nudicaule* are superficially very similar indeed and it is a quite legitimate question to ask why Viguier in 1814 separated the Welsh poppy into a new genus he called *Meconopsis*. The separation turns on two related characteristics of the female part of the flower: the presence of a short style and the absence of a stigmatic disc surmounting the ovary. In layman's terms this means that in *Meconopsis* there is a short 'stem' between the stigma and the inner part of the flower, the ovary, that will develop into a seed capsule. It also means that there is no disc of material forming a projecting collar at the front end of the ovary. Viguier separated *Meconopsis* before any of the Himalayan species had been described, it being a full decade later that Wallich collected material that was described as *M. napaulensis* by De Candolle. It turns out that neither of these characteristics is entirely satisfactory now that we have such a range of different species, and it might be well to understand what taxonomy (the classification of plants) is all about.

There are three basic ways in which botanists seek to classify plants into what, to them, is a logical and consistent order. Taxonomists look for characteristics, usually in dried herbarium material, that will reliably separate different plant material. This may be at the fairly gross level of the family and can put the genera *Primula* and *Cyclamen* into the same family of Primulaceae, or it can divide down to the level of sub-species. Taxonomists are themselves divided into lumpers and splitters who professionally can occupy themselves indefinitely by turning sub-species into species and then changing their minds at a later date. Present fans of the genus *Meconopsis* have much to be grateful for to Sir George Taylor who, more than 50 years ago, lumped a number of species together and made sense of many previous species that were clearly simply extremes of single variable types.

Then there are horticulturalists, who very much look at the way plants grow both in the wild and in cultivation, and what they see as significant in separating a group of plants may be entirely different from those of the taxonomist. The blue form of *M. napaulensis* is radically different in the garden from all other forms of this species and is quite distinct to me (and so it was to the King of Nepal's collector Lal Dhwoj)—but such an idea would offend a taxonomist. The final approach to dividing plants is the evolutionary one looking at relatedness perhaps using modern genetic techniques and is in many ways by far the most fascinating approach. In the end all classifications are artificial and at best compromises.

When Viguier picked on the female flowering part characteristics, it seemed soundly based on the available knowledge. At the time it is likely his colleagues were not over-enthusiastic about the new genus. History, however, proved him right. *M. cambrica* itself is admittedly a bit of an anomaly, but the Himalyan species are really rather distinctive if not to the taxonomist then certainly to the gardener, or plantsman interested in evolution. The problem with taxonomic features is that evolution can go both ways and such is the conservatism of the forces of nature that there are often very few solutions to the same problem. This produces features that appear identical but in reality they are analogous, in that they serve the same function, rather than homologous (as is required by a taxonomist) and derived from the same structure. There was no doubt some underlying reason for the evolution of a style and the loss of the stigmatic disc in *M. cambrica*, but neither were irreversible processes, and some forms of *M. integrifolia* have a capsule as lacking in style as any field poppy (*Papaver* sp.) and several species such as *M. discigera* are classified within *Meconopsis* on the basis of having a stylar disc! Things come and things go. This is a plantsman's book and we bury our heads in the living flowers, but taxonomy is as necessary as a tidy garden shed and we must show some tolerance.

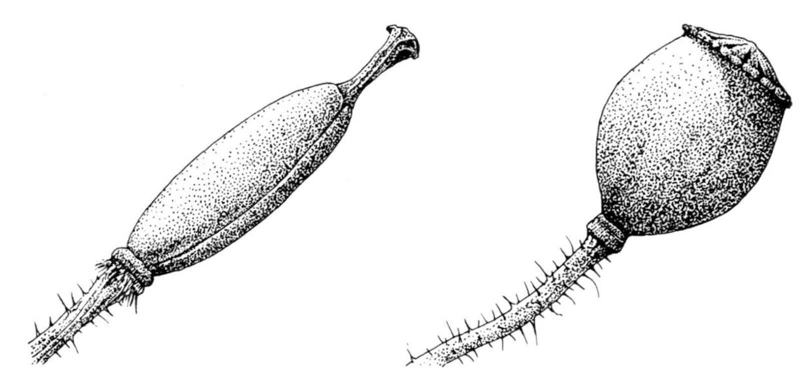

1. Comparison of papaver (right) and meconopsis fruiting bodies showing characters which separate the two genera

We can be a bit dismissive as plantsmen however of two other genera that rear their not so ugly heads, namely *Cathcartia* and *Stylomecon*. *Cathcartia* was a genus described by Hooker well over a century ago and at that time included only *C. villosa*. Taylor placed *C. villosa* and a number of more recently described related species back in *Meconopsis*. There is little doubt that a plant such as *M. villosa* in flower and fruit has a different 'feel' from a classic species such as *M. grandis*. Taylor has thoroughly rehearsed the arguments for the removal of the genus *Cathcartia*. One can summarise this by saying that a few species like *M. villosa* and *M. chelonidifolia* have characteristics which show they are closely related to each other and to the rest of *Meconopsis* but others which show they differ. Taylor reached the conclusion that they could not be sensibly separated and it was better to leave reasonably well alone and include them in *Meconopsis*. One does still see the genus *Cathcartia* used in nurserymen's catalogues which probably suggests they are 50 years out of date rather than grinding a taxonomic axe.

The genus *Stylomecon* has recently been re-included in *Meconopsis* by Stephen Haw but without discussing the reason (1980). *Stylomecon* is now regarded as containing a single annual species from the west coast of North America. We have already accepted that a character such as the possession of a style may be the repeated product of a different line of evolution away from the *Papaver* stock. More simply put, some characters of *Stylomecon* look like *Meconopsis* but they have evolved separately and are not evidence of relatedness. We can dismiss the possibility of this genus being included in *Meconopsis* as well as the other closely related New World genus *Argemone*, and content ourselves with those plants defined more than 50 years ago by Taylor in the monograph that was the basis of his DSc thesis (Taylor 1934).

It has to be confessed that any classification is complicated and that what is necessary here is to look superficially at the classification published by Taylor, which has not been effectively challenged since, and then translate it into a more easily understood form. Taylor's classification is summarised in Appendix II.

The genus *Meconopsis* was divided by Taylor into two sub-genera and most species fall into *Eumeconopsis*. Two or three species are separated off into the sub-genus *Discogyne*. This is based on a purely taxonomic characteristic of a stigmatic disc in the minority group. In practice a species like *M. discigera* in *Discogyne* has the look and feel of many of the other species and you or I would not hesitate to call it a typical prickly blue poppy. We would be much more likely to worry whether the yellow wildling *M. chelonidifolia* was indeed a meconopsis although it is in the same sub-genus as all the classic blue poppies.

Eumeconopsis are a pretty diverse assemblage of species and Taylor divided them into three sections with two further sub-divisions. The first section is *Cambricae* which has the anomalous European species all on its own. The next section is *Eucathcartia* which has two series based on *M. chelidonifolia* and a related species and *M. villosa* and one related species. In practice to the gardener all these, including *M. cambrica*, are yellow-flowered wildling species, attractive in a way but it is difficult not to be patronising when comparing them to their sumptuous relatives.

The final section of *Eumeconopsis* is *Polychaeta*. This is further subdivided into two subsections, *Eupolychaeta* and *Cumminsia*. The former is the group containing all the monocarpic evergreen species that delight us inordinately in winter with their marvellous rosettes of leaves. Taylor separated these into two series with *M. superba* and *M. regia* in one (plus the subsequently described *M. taylorii*) and all the rest in the other. This is based on grounds that need not disturb the gardener.

Cumminsia are more of a rag bag and consist of six series. Two of these series have single species. *M. delavayi* and *M. bella* do not fit too well into other groups though both tend to be perennial as well as difficult. The largest and most distinctive series is the *Aculeatae*. This is the group of monocarpic species which mainly have blue or purple flowers, very prickly deciduous leaves and a totally monocarpic habit. *M. horridula* is the best known of this group. The series including *M. lyrata* and *M. primulina* is related to the *Aculeatae* in general form. The remaining two series are closely related to each other, *Simplicifolia* and *Grandes*, which contain the rest and most of the very desirable species, including *M. grandis*, *M. betonicifolia*, *M. integrifolia*, *M. quintuplinervia*, *M. sherriffii* and *M. punicea*. All these tend to be deciduous and perennial.

We now have a simple gardener's classification of (1) yellow flowered wildlings, (2) evergreen rosette species in the full range of colours from pure white to deep blue but which die after flowering, (3) prickly blue poppies that die after flowering and are also winter dormant, (4) all the rest which can be perennial, usually die back in winter and again show the full range of possible colours that have an intensity and purity that sets them apart from almost all the other genera of plants. This simple classification is reflected in the gardeners' guide to identification (see Key on p. 44).

A really fascinating question about the evolution of the genus as a whole is why *M. cambrica* is geographically so isolated in western Europe. A significant difference is that although the climate of western Europe has been highly variable over the last million years or so, even during the coldest periods there were paths by which plants could escape south. The Himalayas have been subject not only to the same fluctuations in temperature but also to violent geological events as two continental plates collided to form them. It may be that in the Himalayas some ancestor of *M. cambrica* had the potential to evolve favourably, subject to the continuous and radical development of the habitat of the Himalayas. In the west of the range of the same *M. cambrica* ancestor these selective pressures did not exist and the species was maintained relatively unchanged. This sort of speculation is fun but rather fruitless without evidence, but as the genetic basis of the differences between species can be probed ever more effectively in the way genetic fingerprinting is being done in the animal world, then degrees of relatedness may have a sound basis and will tell us much on the origin and evolution of present-day species and genera.

Chromosome analysis already tells us some things about relatedness and evolution. The genus *Meconopsis* has been well studied by J.A. Ratter (1968).

He worked using 17 species of *Meconopsis* growing at the Royal Botanic Garden in Edinburgh. Most organisms have a double set of chromosomes, one from each parent. The actual number of chromosomes in each single set

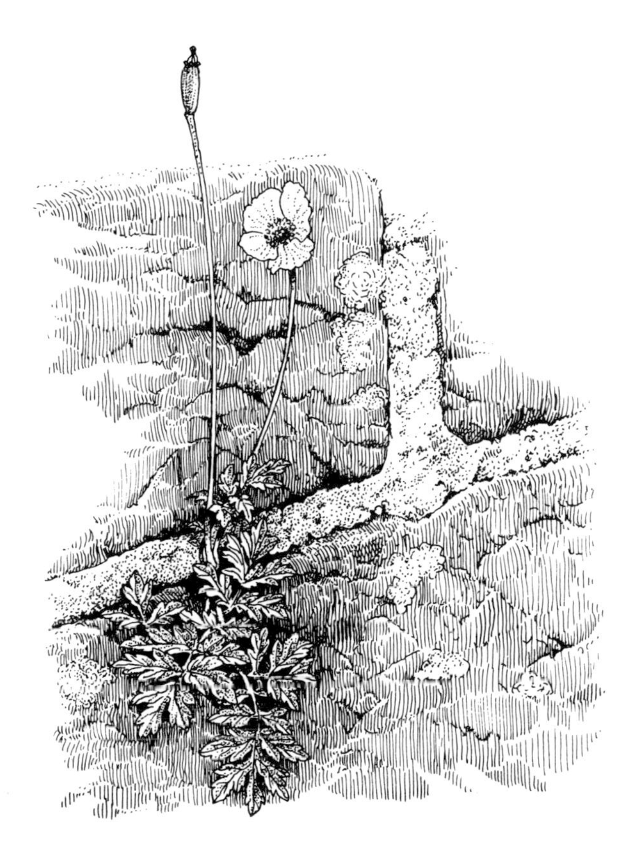

2. *M. cambrica* self-seeded in a roadside wall in Wales

may vary in plants but in all angiosperms (flowering plants) it is held to be between 6 and 12 in the primitive state. In *Meconopsis* this basic set is $x = 7$. Vegetative cells of plants in this primitive condition would thus contain two sets and thus $2n = 14$ (x and n are the symbols given by those who study plant genetics). Evolution often sees an increase in the number of sets of chromosomes, a tetraploid plant having four sets. Polyploid plants can have quite large multiples of the original diploid condition of the primitive species but it must be noted that even plants of the same species in the wild can have different numbers of sets of chromosomes. Chromosomes can also be lost during evolution so that the polyploid may not be an exact multiplication of the diploid state.

The original state in vegetative cells in *Meconopsis* would appear, as shown above, to be the diploid state (2n) is 14 (i.e. 7 pairs). *M. cambrica* is tetraploid at 2n = 28. The rosette-forming species and the relatives of *M. horridula* are octaploid of this at 2n = 56 and the perennial species are more variably polyploid at higher numbers. In general, the more polyploid a species becomes the less effective a mutation of a single gene on a chromosome is, since with a number of copies of each chromosome the effect becomes diluted when translated into some feature of the plant. The implication of this is that highly polyploid plants (which can occur at any time during evolution) have little potential for further evolution but are not necessarily of recent origin. A diploid plant has much greater potential for further evolution. For example all living *Magnolia* are highly polyploid and the genus has little potential for further evolution. *M. villosa* is unaccountably peculiar with 2n = 32. Appendix III lists known chromosome counts.

The knowledge of chromosome numbers can be of some help in understanding hybridisation. Plants with different chromosome numbers may be less likely to produce fertile hybrids. Plants can become isolated in ways that generally do not apply to animals. Species can arise, with a separate gene pool, when groups of plants become isolated and gradually evolve into distinct populations. They may eventually become so different that taxonomists would recognise them as separate species. If however the barriers that separate them are removed, say when brought close together with many other species in the unnatural conditions of the garden, they may still cross-fertilise and produce a new species—effectively as a hybrid. In simple terms it is not necessarily a part of the botanical definition of a species that it should be infertile when crossed with other species. The offspring of many are sterile but some inter-species crosses are fully fertile and may eventually breed true. This is indeed one way in which new species can evolve. The continual violent change in the Himalayas may well have made this a significant factor in *Meconopsis*. Some hybrids are fertile in meconopsis, such as *M.* × *beamishii* (between *M. integrifolia* and *M. grandis*). Other crosses are sterile, such as that between *M.integrifolia* and *M. quintuplinervia* (*M.* × *finlayorum*). Some are very difficult to produce, such as *M.* × *auriculata*, between *M. violacea* (an evergreen monocarpic rosette species) and *M. betonicifolia* (a perennial deciduous species). It may be in these latter types of hybrids where the chromosome numbers of the parents are very different that complex multiplication of the number of sets of chromosomes is taking place to produce the hybrid. What should concern us however is that it shows the potential for incorporating desirable genes for perennial behaviour in monocarpic species.

Meconopsis have the purest red, the softest pink, the most exquisite blue, the clearest yellow, the most faultless white and the most imperial purple of any genera of plants and, in most species, very few intermediate colours. Colour in plants is due to a number of genes which code messages for particular pigments, anthocyanins are associated with red and blue colours and flavones with yellow and ivory. There are also genes which suppress pigment and produce white flowers and those that enhance or bleach colour produced by the pigment genes. There can also be interaction between genes. In some polyploid plants, for example a tetraploid, there can be a dominant gene on only one chromosome of the set of four, or on two, or even on three or four. This situation produces a stronger and stronger colour effect. Although most species of *Meconopsis* are apparently polyploid, one rarely sees this effect, which is surprising. We know from hybridising that crossing a blue species with a yellow species produces a cream hybrid (hybrids of this type occur in the wild, such as *M.* × *harleyana* which is Kingdon-Ward's 'Ivory Poppy' between *M. integrifolia* and *M. simplicifolia*). There are however no cream-flowered species (except possibly some forms of normally blue *M. discigera*). It is even more surprising to find that in a number of blue species, such as *M. horridula*, there are yellow forms but no reported cream intermediates. *M. integrifolia* (the parent of many cream hybrids) is always a perfect yellow, *M. punicea* a perfect red and *M. latifolia* a perfect blue; such purity of pigmentation is intriguing. Hybrids between red and blue produce a muddy pink colour but this can also occur because of conditions of cultivation. In *M. napaulensis* crosses between red or pink and yellow can produce almost chimera-like flowers with patches of both colours or sometimes the orange/yellow gradations associated with the more vulgar type of hybrid tea rose. The explanation for this may have a very complex genetic basis as has been shown for the 'broken' colours of antirrhinums. The isolation of the pigment types in meconopsis is an interesting problem and the genetics of this may well repay study, but we should just be thankful for the extraordinary purity of colour.

1 *Places and People*

Places

In the back garden, I sometimes say, I grow Himalayan plants. This is rather as though the Himalayas is a rather good sort of garden centre down the road. It is I suppose a penalty of having been born in a small island that I had little real appreciation of the scale of many of the great geological features of the world until I visited them. Having flown for hours in a jet across Australia without much change in the landscape below, and more recently having 'done'

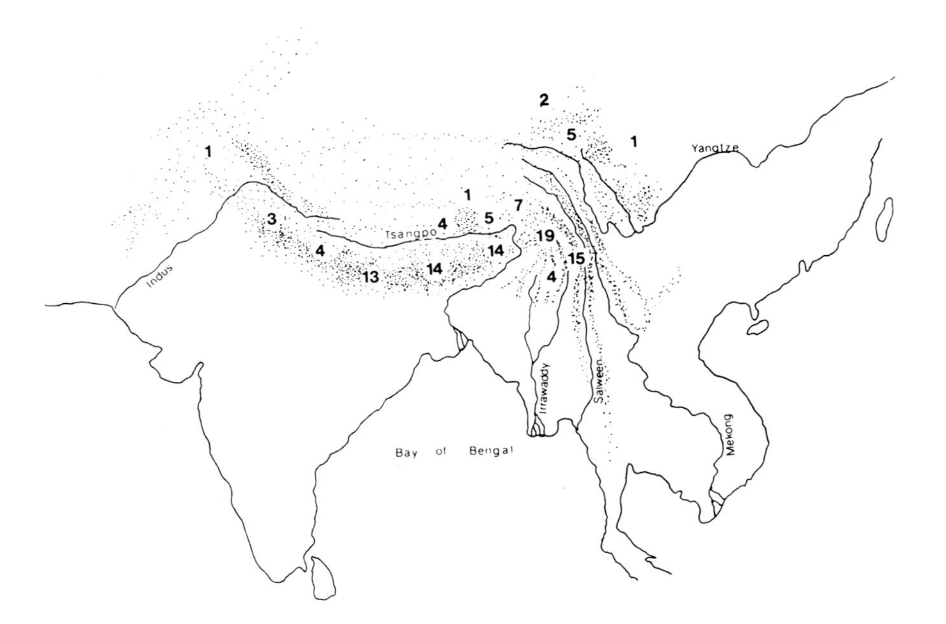

3. Species distribution of Himalayan species (*M. cambrica* does not occur in the area shown)

the Rockies and then discovering I had only circumnavigated a pin-point on a map, I began to realise the absurdity of calling anything a Himalayan plant in the sense that it typified something.

The true Himalayas (see Figure 3) stretch for something like 2,700 km (1,700 miles) but they do not run directly east-west but are tilted so that the western end reaches 36°N and the eastern end only 27°. Even within this north-west to south-east tilt, the main backbone forms a flattened S-shape. In the Hindu Kush and Arunchal Pradesh (Assam) at either end they run north-east with an east and south-east component through Kashmir, Nepal, Sikkim and Bhutan. This distance covers a massive 63 degrees of longitude. The difference in latitude is equivalent to that between the United Kingdom and the Mediterranean and this without the climatic effects of the mountains. The main range varies in width between 80 and 320 km (50 and 200 miles) but one has to bear in mind that the foothills on the Indian side would be major mountains in much of the rest of the world and the Tibetan plateau behind stretches north at a steady 4,500 m (15,000 ft) with higher ground of permanent snow and glaciers.

Understanding the sheer size of this mountain range is not enough, however, because the rate of change is often exceptionally violent with walls of mountain rising 6,100 m (20,000 ft) in a few kilometres. It is of course the land of the highest mountains and deepest gorges, geologically recent and subject to the awful power of wind, rain and ice that seek to level anything that protrudes above the thin skin of the living planet.

It is deluged by enormous volumes of water brought by the monsoon and as glaciers and rivers this is returned to the Indian Ocean by valleys which run in every direction of the compass. Two of the great rivers of the Indian continent, the Indus and the Brahmaputra, rise close to each other on the Tibetan side of Nepal and flow west and east respectively for hundreds of kilometres before finally breaking through the barrier. A look in detail at a map of any part of the range will give an instant impression of the immense complexity and rapid change of the geography.

This is however a simplistic view of the Himalayas in terms of what we call Himalayan plants and of meconopsis in particular. Strictly speaking the Pamirs in the north-west are part of the Himalayan range but are of no interest to our account since no species of *Meconopsis* have been recorded there. The mountains of that part of China abutting what was Tibet, as well as Burma and Arunchal Pradesh, are not part of the Himalayas geologically but certainly are floristically and are of vital interest to us.

The Himalayas really end when the Brahmaputra, disguised at this point as the Tsangpo, escapes the Himalayas by passing beneath the great Namcha Barwa. Thereafter in rapid succession are three great rivers, the Salween, the Mekong and the Yangtse, all running north-south, although the latter does a great double bend before proceeding east. These three rivers are within 160 km (100 miles) of one another and run parallel to great ranges that run north-south and these too are affected by the monsoon sweeping up from the Indian Ocean. To the north these ranges join the cold bleak plateau of Tibet and they run well south of the Himalayas almost to the Tropic of

Cancer so that for all their alpine nature some meconopsis are virtually tropical plants!

Finally the arid Tibetan plateau also interests us because a few species of meconopsis have struggled to find a habitat and at least one species is endemic to that region.

The origin of the Himalayas is significant floristically both for the soil types and the timescale. The upthrust of the Himalayas over the last 15 million years has been caused by two continental plates colliding. Associated with this has been volcanic and earthquake activity, the latter of which is continuing. The sea-bed of past ages is now the mountain tops and marine fossils occur in limestone rocks approaching 6,100 m (20,000 ft). The same elevated limestone beds occur in the north-south ranges of western China. In an area that is of the greatest importance in the evolution of the genus *Rhododendron* this presence of limestone was, and to an extent still is, an enigma in terms of our cultural practice with these plants. It should dispel however the idea that Himalayan plants are all naturally acid soil loving.

There is evidence that the last million years saw the Himalayas thrust another 3,000 m (10,000 ft) up and are still rising. This is almost as yesterday in terms of biological evolution and this rapid change will clearly have caused tremendous pressures of selection on both flora and fauna. It is of course the ideal situation for 'playing' at evolution. In the first place truly tropical plants flourish close to habitats so inhospitable that no vascular plant can survive and this within 2–3 km (a mile or two) of each other. The churning of the rocks by tectal plate movement and the grinding by ice and the leaching by water assured very variable soil conditions over equally short distances. The regular watering from the monsoon and the high winter deposition of snow provided a regular, if not regulated, moisture supply. This too was highly diverse in extent, even very locally, because of the shape and the height of the mountains. The very proximity of these habitats to one another was a vital ingredient because very short-term deteriorations or ameliorisations of the climate, typical of the recent ice ages, could be easily escaped by moving up or down, or along, a ridge. It is a marvel, but not a surprise, that such genera as *Rhododendron*, *Primula* and even *Meconopsis* have evolved to exploit such a range of habitats. This does not entirely explain the floristic brilliance of these genera unless it was caused by inter-generic and inter-specific competition for pollinators putting selective pressure on to the development of large flowers; but then if we understood everything life would be much less fun. Fortunately science does not solve problems, it just makes the questions more difficult!

This then is the habitat of all the meconopsis, bar the curious *M. cambrica*, but however significant in our quest to cultivate the genus, the climate can only be described in general rather than in detail. It is disappointing that even after reading all the many accounts of plant explorers one is little wiser, so varied is the climate, soil and exposure over very small distances.

Travelling over a high pass may alter the flowering season by two months.

It is possible however to apply a few broad brush strokes to the corrugated canvas of the Himalayas and adacent China. From June to September the climate is controlled by the Indian monsoon regime. It is not a continuous rain but shows in some areas an oscillation at about 10-day intervals that may be associated with the massive anticyclone over Tibet or the monsoon regime itself. A five-day cycle may also be imposed especially in east Nepal by the passage of depressions across northern India. Finally on top of this there is a daily effect associated with the heating of the land by day and an up-valley movement of moist air. The intervening days can be quite dry, at least for part of the time, apart from local convective showers. It may well be very significant for the proper pollination of so many flowers during the monsoon that these dry spells do occur. The rain also falls in an altitude-dependent way, often leaving the valley floors much drier than the ridges and cols. This is due to oreographic (lifting of the air mass due to the height of the mountains) and thermally induced convection; radiative heating below the snowline will also be significant.

From autumn until the end of May the whole Himalayas is subject to the influence of the sub-tropical westerly jet stream (high altitude wind). In winter at the height of 9 km (5.6 miles) this reaches an average of 90 kmh (55 mph). This wind decreases in spring until eventually it is replaced by easterlies associated with the Tibetan anticyclone. The precipitation associated with the high altitude westerly airflow increases with latitude and altitude and is much more significant in the west. This winter snow in the NW Himalayas is of the greatest significance as snow melts in the summer. In the Karakoram for instance there may be a few centimetres of rain annually in the valley bottoms but several metres of snow on the surrounding glaciers and mountains. The actual precipitation in this part of the Himalayas is much more dependent on local conditions than any general factor.

The Himalayas act as a varyingly effective rain barrier to the monsoon and the inner valleys and in places, such as Leh in Kashmir, it may be almost totally effective in drying out the humid air being lifted and squeezed over the mountains. It is dependent very much on the inclination and height of particular ranges, and at the Chinese end the Mekong valley is arid because the humid air is not sufficiently saturated to have moisture squeezed out of it by the modest mountains of that divide but subsequently this same north-east moving air mass produces substantial rain over the Salween and Yangtse ranges. Two or three km (1–2 miles) can make an enormous difference to levels of precipitation and the plants we are interested in may be confined to an area the size of a football pitch, thus extreme detail is required to make any sense of conditions.

The area to the north of the Himalayas, which is effectively the Tibetan plateau, is subject to summer precipitation associated with large cumulonimbus cells (thunderstorms to us mortals) but very occasionally there have been monsoon effects recorded. In winter this area is subject to bitter winds from the north and often the plants are not protected by snow cover and are subject to continual sub-zero temperatures. The escape from the monsoon climate to the dry continental climate is more gradual in the north-south ranges of western China. The Min Shan range at the north really acts as a barrier against winds from Mongolia. The mountains of Omei Shan, Wu Shan and Wa Wu Shan in the area north-east of the Yangtse, called the wilderness, are influenced by the monsoon but are really islands of great floral wealth rather isolated from the rest of the mountains.

People

This curving backbone of the Himalayas and the ridges of adjacent China are in the main very inaccessible, but the great floral wealth has excited the imagination of plant lovers and stimulated more than a century and a half of botanical exploration. The inaccessibility has not only been physical but also political and to this day many areas are still largely unexplored for the latter reason. All the main areas have however been subject to fairly intense search during one period or another and some have been very well documented.

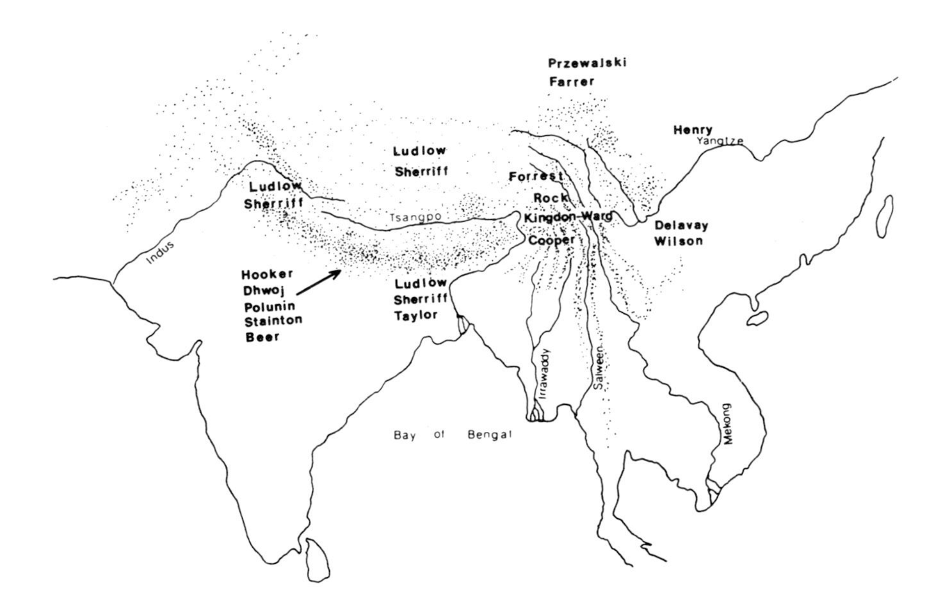

4. Area covered by major collectors

The western end of the Himalayas sees the meeting of the flora of the Middle East—typified by bulbs adapted to survive a hot dry summer and having only a fleeting spring to flower in—and the monsoon flora of the Himalayas. In some ways it has the worst of both possible worlds with only the outliers of both. The most westerly species, other than *M. cambrica*, occurs here but is poorly documented and was described as *M. neglecta* by Taylor from poor specimens. It is surprising to find that both Ludlow and Sherriff spent some time in the west between their eastern Himalayan adventures but, apart from the occasional article by Ludlow, they recorded relatively little in print although they collected more than 3200 specimens in this area. The region of Kashmir has had some tourist traffic recently and the habitat of *M. aculeata* has been visited by a number of keen gardeners of the current generation.

The flora of Nepal was hidden from western view for 100 years after Sir Joseph Hooker's visit in the middle of the last century until it was re-opened to the West in the 1950s; the expeditions of Stainton, Sykes and Polunin and later of Stainton, Sykes and Williams described many exciting plants, particularly marvellous colour forms of *M. napaulensis* and *M. regia*, the progeny of which, somewhat mixed up, still grace our gardens. Oleg Polunin and Adam Stainton much later produced the excellent *Flowers of the Himalayas* (1984). More recently expeditions led by Binns and the late Len Beer explored further areas. It must not be assumed however that because Western botanists were denied Nepal for the first half of this century that plant exploration did not occur. Maj. Lal Dhwoj, who had training at the Darjeeling Botanic Gardens, was appointed to explore for plants and must be credited with introducing three new species: *M. gracilipes*, *M. regia* and *M. dhwojii*, the latter obviously named after him. He died while engaged on this task which was then continued by Professor Sharma appointed by the Indian government.

Sikkim was explored at least in part by the series of expeditions of Ludlow and Sherriff both before and after the Second World War. They were accompanied by Sir George Taylor and John Hicks (the expedition medical officer) on later expeditions into Bhutan, where they explored systematically the high passes towards Tibet almost into Burma and China and to the edge of Kingdon-Ward's territory. Ludlow and Sherriff spent time during the Second World War in an official capacity in and around Lhasa and added to our knowledge of the few species of meconopsis that grow there including the endemic *M. torquata*. These expeditions introduced much seed as well as the fabulous pink *M. sherriffii*. They produced quite excellent colour photographs in the field of many species. They also pioneered the flying home of plants, especially in attempts, sometimes successful, to introduce primula species with short viability seed. It is a pity in some ways that neither Ludlow nor Sherriff wrote more, and growers of these plants must be especially grateful to H.R. Fletcher who wrote up the Ludlow and Sherriff expeditions in the absorbing book, *A Quest of Flowers*, as well as publishing some of their colour photographs (Fletcher, 1975). It must not be forgotten that Sherriff and his wife Betty (who accompanied him on later expeditions) retired to garden just across the Tay from where I write. They successfully grew many of the most difficult plants they brought back at their garden at Ascrievie and I am lucky enough to have been shown around the garden by Mrs Sherriff. The Rentons developed the world-famous Branklyn Gardens by the side of the river Tay in Perth (which are now run by the National Trust for Scotland) and with the Knox-Finlays at Keillour Castle, not too far away, pioneered the techniques of cultivation which form the basis of our current abilities.

Frank Kingdon-Ward was active for many years in the first half of the century in 'the bit on the corner'. This is the section between what was Assam (The North East Territory), Burma, SE Tibet and China. This is close to the presumed epicentre of *Meconopsis* development and his work is very important. He wrote a whole series of charming books, now fortunately being reprinted, and although they tend to concentrate rather much on anthropological detail (Kingdon-Ward certainly had an eye for a pretty girl and clearly collected native pin-ups) careful reading is highly informative. There are really first-class speculative accounts on the climate and geology of the area which shows him to have been a most objective and thoughtful scientist. He has been eclipsed, for different reasons, by George Forrest and Reginald Farrer and he is much underrated as a plant explorer.

Forrest was I suppose a magical figure and like all great Scotsmen has had plenty of his kin to champion him. Farrer was larger than life, a truly unforgettable man by virtue of his unutterable brilliance at what would now be considered 'going over the top'. Forrest, who eventually died in harness just as he was about to retire, overlapped somewhat with Kingdon-Ward, and there is no doubt they were all jealous of 'their' territories. They even met in the field, as did the American collector Joseph Rock, who was also active in this area. Forrest was immensely thorough and totally painstaking in his desire to see good plants introduced. Life was undoubtedly hard, lonely and most certainly dangerous for collectors like Forrest who spent long months in the field. Forrest in particular collected literally kilograms of seed of many of the most desirable species, and perhaps the sheer profusion of material was why some of it appeared unresponsive in cultivation. As the century progressed the economic climate in the West was changing from an era of vast estates and the staff to run them, and it is only now with many more amateur plantsmen with more spare time that we are ready to receive treasures such as Forrest's, if they can be re-collected. The one regret is that Forrest wrote so desperately little of his work while still alive, maybe because he thought it a job for the retirement he never achieved. Forrest and Farrer are commemorated by medals for the most prestigious plant at shows of the Scottish Rock Garden Club and the sister organisation, the Alpine Garden Society, respectively.

E.H. Wilson collected to the east of Forrest, more clearly into regions of truly Chinese flora, and fortunately he wrote up at least some of his journeys.

Farrer initially collected to the north of Forrest at the top end of the north-south valleys of the Salween, Mekong and Yangtze. It is a much drier area and, although Farrer has some typically lyrical descriptions of such beautiful species as *M. integrifolia*, *M. punicea* and *M. quintuplinervia*, he could not really disguise the fact that his Stone Mountains were more affected by cold dry Siberian winds than by the humid warmth of the monsoon and floristically were much poorer than the area where China meets the Himalayas. Forrest and Farrer both ventured into south-east Tibet, both to their peril; then, as now, the natives objecting to Chinese rule. The area described as the wilderness, because of the very wet climate but including isolated mountains such as Omei Shan, was briefly explored.

There is no doubt still much to do in the land of meconopsis. It may not be dramatic new species, only variations on what we now know, but we have hardly started yet on collecting sufficient information. Learning to grow these plants requires dedication, and the words of the people who went to these remote places can often be the inspiration for cultural experimentation—growing plants is a science as well as an art.

The late E.H.M. Cox of Glendoick in Perthshire, who accompanied Farrer on his last journey into the area north of Burma, wrote at length on the plantsmen who explored China (Cox, reprinted, 1986) and he too lamented the poor detail about the habitats, climates and soils of the areas. He admits however that the energy to self-motivate oneself to the physical tasks of a plant collector, the need to provide for the bodily needs of oneself and one's native servants, to endure the mental burden of being alone as well as writing it all up in sufficient detail to please a scientist (and with the literary skill to make it readable) was asking a bit much even for the superhuman. Today we are more generous with such expeditions and expect less from the muscles and mind of a single man and send in teams of experts. This is another phase and a new sort of knowledge, less romantic perhaps but just as exciting. Sinclair and Long of the RBG at Edinburgh are active in Bhutan and Ron McBeath, David Chamberlain (both also from Edinburgh), E.H.M. Cox's son Peter of Glendoick rhododendron nurseries and Chris Grey-Wilson (from Kew) among others have worked jointly with Chinese botanists.

2 *Habitats*

Garden habitats

All species of meconopsis appreciate rich feeding and although some will tolerate a poor starved soil and even seed themselves into it, the resulting plants are pale shadows of what they could be. Some of the very high altitude forms of *M. horridula*, and perhaps some of its relatives such as *M. speciosa* which are highly desirable but not in cultivation, might become unduly soft if well-fed and succumb in winter.

There is then the vexed question of whether they are lime-tolerant or even lime-loving. Meconopsis are regarded as classic plants for the peat garden, certainly by the current generation of growers. They are plants of forest glades, meadows and alpine screes and although they do undoubtedly occur in the equivalent of the acid peat moor in the Himalayas this is the exception rather than the rule. Many of the Chinese areas where meconopsis grow are limestone (admittedly dolomitic) and I suspect this is true of other areas of the Himalayas. I have grown 20 species on an alkaline soil and do not believe acid conditions to be at all critical, with one or two exceptions.

I have gradually come to the conclusion that in some species (but by no means all) the perfect blue colour is compromised by an alkaline soil. This is so in *M. betonicifolia* and particularly so in some forms of *M. grandis* but it is most clearly shown in the hybrids *M.* × *sheldonii*. These can turn an even light purple without a trace of blue under alkaline conditions. There are other factors such as temperature which may complicate this. The dark blue of good forms of *M. horridula* and the light Cambridge blue of *M. latifolia* are not affected by soil pH. The one other species that concerns me is *M. superba* since the even silver-green of the leaves does sometimes show a yellowing typical of calcifuge plants in alkaline conditions although it may well be that this is a drought effect. There is certainly no need to refrain from attempting any species simply because the soil is alkaline, as long as it is full of organic The species most associated with limestone areas is *M. delavayi* but R.D. Trotter of Brin in Inverness, who grew this difficult plant so well, found the addition of lime to the soil to be insignificant.

Meconopsis can be divided into rich scree and rich meadow types. The difference between these is simply that the former is probably in greater sunlight, on a slope and with more stones in the soil and the latter has a soil richer in humus. Peat beds for meconopsis should not surprisingly be largely composed of peat. This substance provides an open, well-aerated soil which is usually acid but it is a fairly sterile medium and exactly the opposite in terms of feed from what meconopsis require. There is a fundamental misunderstanding about peat beds and the use of peat in general. It can be used as a surface mulch where, if an appropriate depth is used, it can act to conserve moisture. In general the huge surface area of a moss peat uncompressed on the surface is going to evaporate substantial amounts of water; sterilised garden compost is much better. It is also a weed suppressant but the very fact that weeds find it inhospitable is an awful warning about the lack of nutrients in it. Thus digging peat into the garden as some sort of food, as is done in so many different garden situations, does little beyond temporarily aerate the surface layer of the soil and perhaps equally temporarily increase the water-holding capacity. Peat does of course contain a great deal of organic matter but it is in a very inaccessible form, and nutrients have to be released from the soil before soil organisms can begin to work on breaking down the peat and releasing its potential goodness. This is a slow process and if peat is added annually this will merely re-lock up what may just be beginning to be released from the previous year's application.

Peat is however of great importance in creating the basic soil structure, especially if damp acid soils are not a natural feature of a planting site. The best sort of peat is probably a coarse moss peat. It is likely to be expensive and the amount required to make a really good peat garden is prodigious since one probably needs to mix a layer 45 cm (18 in) deep into the soil in the first place. The alternative is to use a sedge peat. This is more highly compressed and in general is probably a much better buy if bought by the cubic measure since you purchase much less air for your money! It may well be less acid and as producing acidity is a primary consideration for blue poppies that are truly blue it may be worth finding out the pH of the peat before purchasing it in bulk. It must be confessed that if the soil is naturally alkaline and an acid woodland soil is created on top of this, eventually the ground water will be drawn up and the soil pH neutralised unless the peat is taken to great depth. It is possible to create a panned layer of peat by applying a thick layer 30 cm (1 ft) deep and stamping this down as hard as you can before using soil, more peat, leafmould and compost to create the rich acid bed on top. This will probably last a life time and doing this over a few square metres could provide ample space for many treasures.

Comparison with habitats in the wild

The absence of acid or peaty soil should not, however, put one off growing many of the genus *Meconopsis*, nor is it necessary to contrive such habitats. They can be grown almost anywhere provided one basic requirement is understood and that is they need feeding. It may well be that in an alkaline soil the blue of *M. grandis* and *M. betonicifolia* have washes of purple but to the layman they will still appear fantastic; only if you visit one of the classic cultivation sites in Scotland will you feel twinges of jealousy or frustration.

The cultivation of particular species is dealt with individually in Chapter 5 and what follows is a general guide to the two basic types of cultivation that are necessary. In the wild, habitats can be separated into two basic types, meadows and screes. In meadow situations meconopsis occur in high valleys with a good depth of alluvium that is rich in the decayed humus of thousands of years. There is one report of a soil analysis from such a situation and the organic matter in the soil was 30 per cent of the total. These areas would be rich in many herb species and with an amount of scrubby cover. They may occur in clearings below the top of the tree line of in meadows above the upper limits of tall tree species.

There are many wild grazing animals in the Himalayas as well as domesticated animals, and large concentrations of dung are dropped in those areas where they stand and ruminate. The edges of clearings have many advantages of shelter, perhaps from annoying insect species, shade from the heat of the day and the safest position from predators such as wolves since retreat into thick cover is combined with good visibility. These areas, rich in organic matter with a high nitrogen content, are a natural feature for many species of meconopsis to exploit. There are numerous accounts of the excellent quality of primula and meconopsis species in the vicinity of shepherds' dwellings for exactly the same reason.

The habitat of scree-dwelling species is different since there will be a lesser accumulation of organic matter, although even here flat platforms at high altitudes may yet again be especially favoured by grazing animals which use the high areas well above the tree line for feeding in summer, often escaping flies by doing so. Food and nesting material may be collected by small rodents and be an accumulated and buried source of nitrogen. In these high altitude screes there is likely to be continual leaching downwards of mineral salts and these are constantly replaced from above. These areas would initially be nitrogen poor unless a high incidence of electrical storms fix significant amounts of atmospheric nitrogen, but eventually herbs adapted to high salt conditions would decay and increase the organic matter.

This knowledge of the two basic habitats needs to be adapted to a garden situation.

Meadow species cultivation

Aiming for perfection

It is essential for perfect cultivation to have adequate summer rainfall and preferably cold drier winters. There is little doubt that the Scottish climate is very good for meconopsis. Similar conditions will apply in northern parts of coastal North America (both east and west coasts). There are problems however even here in Scotland, since the west coast is too winter wet and the east coast is too summer dry and the climate unpredictable in both. The most important thing to realise is that a temperature above 27°C (80°F) is exceptional and that although 16 hours of sunshine a day in mid-summer is possible the sun does not have the burning intensity of even the south coast of England. It is necessary to have mature trees surrounding the planting site and the majority of these should be deciduous with conifers only as an outer windbreak. Shrubs such as rhododendrons are necessary at lower levels to reduce ground wind speed. This is as much as anything to ensure that summer humidity is maintained within the glade rather than constantly blown away. Marjorie Brough writing before the war described exactly this habitat in a Hertfordshire garden and she said no more unlikely overall climate existed in Britain. There is little doubt that her garden achieved perfect cultivation for meconopsis.

In many places with a dry climate the potential to create a woodland glade only needs attention to the soil if all the major tree and shrub factors are already in place. It took me ten years starting from scratch to make a woodland glade in an open field on the most barren part of the east coast of Scotland.

It is always easier to contemplate raising the humidity and watering things than drying things out. Areas with a high winter rainfall may be difficult to adapt to perfect conditions.

The soil must be enriched and should be naturally acid. Producing an acid soil is possible with peat, as described on p. 17, and an open texture is essential. A heavy clay soil might grow plants well in summer but there would be great risk of unacceptable winter losses. The best material to enrich the soil would undoubtedly be masses of well-rotted leafmould. In a natural woodland setting all that is necessary is to collect up the leaves on an annual basis and have three heaps, using the compost in rotation after three years. The alternative is to bring in leafmould from a site elsewhere but it is undoubtedly a precious commodity. Another solution, which is probably as effective but not quite so aesthetically desirable, is very well-rotted cow or horse manure or garden compost. I use composted seaweed because it is available but this material will undoubtedly be alkaline from the many encrusting animals with limey shells found on seaweed and this should be avoided with the blue perennial species and hybrids. There is one golden rule with all this material and that is it should be dug in and not used as a top dressing.

As one becomes more familiar with meconopsis one is struck by the huge volumes of soil that are penetrated by the fine fibrous roots. These will descend 45 cm (18 in) even in species without a tap root and twice that if they do. This makes it clear that they are not adapted like so many classic peat garden plants to a leaf litter deposit exploited by surface roots. This poses serious problems for maintaining meconopsis in a mixed planting as discussed in Chapter 4 on plant associations. The monocarpic species and in particular the evergreen rosette-forming species require planting over soil that has had substantial amounts of organic material incorporated. This might be as much as a barrowful per square metre (yard) initially but as soil organic matter builds up this could be reduced to half. The surface should still be an open peaty mixture and if there is any worry about surface water drainage in winter, the incorporation of up to 20 per cent coarse grit would be valuable—and a slope is better than a flat bed. These plants will then need no further feeding until after they have flowered perhaps three, four or even five years later when the whole soil must be re-enriched. The rosette-forming species will produce open and rather lax rosettes in heavy shade but these are still very attractive especially in winter when much else is bare. The flowering stem however becomes unduly elongated and exceptionally may reach 2.5 m (8 ft) and the cymes of flowers from the leaf axils may extend to 60 cm (2 ft) and droop like a weeping willow. Some forms tend to do this at the best of times and it is a poor imitation of what they should be. There is no doubt that as much sun as they can stand, compatible with humid woodland conditions, is desirable to produce really compact flowering plants. The monocarpic species that are winter-dormant like *M. horridula* often look best on a sunny scree but will grow well in rich meadow conditions and reach substantial size. They will become unduly drawn in heavy shade and poor forms may produce great expanded flowering stems with few flowers. They would do well in a drier area of a woodland glade.

The perennial species, particularly the classic *M. betonicifolia*, *M. grandis* and the hybrid *M.* × *sheldoni*, need exactly the same rich soil. They too should be planted into a deeply dug and enriched bed whether the planted material is grown from seed or vegetatively propagated. The plants will fairly quickly occupy about 30 sq cm (1 sq ft) each and will gradually expand beyond that. They will be at their best at about three years and will continue in really good conditions for twice that time before beginning to exhaust the soil and deteriorate. Initially, when stocks are being built up, especially vegetatively propagated forms, the plants can readily be divided after three years and a new bed made up. Eventually however one has to remake the bed or, better, plant in quite a different site.

There is no reason why almost all species should not thrive under these conditions providing the extra care required by a few is recognised and applied. *M. bella* is an exception since this very difficult plant is crevice dwelling and may prove to be adapted to very specific growing conditions. The high altitude forms of *M. horridula* and the related species from western China (such as *M. lancifolia*) that have not been in cultivation or have been quickly lost seem likely to need more specialised habitats.

The test of perfection is simple. If *M.* × *sheldonii* has flowers of such an intense blue-green colour that although you see it every year you still cannot quite believe it exists and at the same time every leaf including the oldest and most basal leaves are a fresh unblemished and even shade of green, then you have achieved it. Less than perfection is a whisper of purple in the bloom or a brown edge to a single leaf. Perhaps such perfection only occurs in the eye of one's mind but I have seen it the garden at Balruddery overlooking the silvery Tay.

Highly commended growing conditions

This is what trying hard should achieve and, even if moderately well done, the blue poppies will stop all but the most philistine in their tracks. It is what one may have to be content with in dry and unsheltered places where there are not enough years to grow the shelter belt, or where summer in particular is dry and windy. Some wind shelter is required and this may have to be artificial, even in the long term, using larchlap fencing. Variable shade too is equally essential and although this could be the north or west side of a building, planting a rapid-growing deciduous cover, particularly birch or even eucalyptus, is much better. It is advisable if dry summers are likely to be the rule, to incorporate mist nozzles at least a metre (yard) high over the middle of the bed. The soil should have an open moss peat incorporated to ensure a humidifying summer surface and a rapidly drying winter micro-climate. Great attention must be given to digging in plenty of rich organic material. The amount of water required even in drought conditions by this type of bed need not be great. If the bed has been made properly the large amount of organic matter at depth will provide water for some weeks; but for many species, however wet the soil, they transpire more water than their roots can provide in dry summer conditions and flagging foliage is inevitable. The only cure for this is morning and evening misting and only a few litres (pints) of water will wet all the surfaces. The evaporation of this from the open peaty surface is slow enough to produce a humid micro-climate for at least part of the day and all the night from two brief applications. There will be a tendency to mildew and the outer leaves will scorch but providing the food below is rich most species will survive. It is certainly possible to grow excellent specimens of many monocarpic rosette species, and robust forms of *M. betonicifolia*, *M. grandis* and their hybrids, and to maintain a good vegetative increase in the latter and ample seed from the former.

Good climate, poor soil

It is not easy in some places to obtain organic material, especially in many city and suburban areas where so many people now garden but where the climate is reasonably wet in summer and dry in winter. This is so of many garden situations in most of Britain, some coastal regions of northern North America and much of northern and middle Europe. There are two solutions to this problem, one better than the other. It is usually possible to buy peat, which is clean, light and pleasant to handle, even right in the middle of London. It is of course extremely expensive but in many urban gardens lack of space is the primary consideration and a few square metres (yards) may be all that could be allotted anyway. The more peat that can be incorporated the better and it is surprising how rapidly even 15 cm (6 in) of peat sinks to be a thin covering: buying twice as much as you think you can afford is probably a step to be encouraged. If the ground is badly drained then a raised bed is to be preferred; although peat blocks look nice they are difficult to obtain and dry out appallingly unless constantly cosseted with the sort of attention the plants deserve. A good sound edging of bricks or stones mortared together in a wall 15–30 cm (6–12 in) deep is ideal. The plants can then go in and feeding applied on a properly scheduled basis.

The easiest food is without doubt a balanced resin-based slow release fertiliser. This releases nutrients particularly in response to water and temperature and a spring application will probably last the season. A second application in early July, especially with large rosette species that grow well into late autumn, is better still. This is expensive but works extremely well. The alternative is a simple liquid feed such as tomato fertiliser (more K than N) on a fortnightly basis, or a couple of well-spread applications of a granular fertiliser that has a balance of nutrients applied in early spring before top growth starts and again after flowering in the perennial species. They may need watering or preferably misting in a hot summer spell. One has to accept however that even peat may not be available or affordable. Species like *M. paniculata* which are very easy from seed, or *M. betonicifolia* which is widely obtainable from the trade as well-grown plants, will survive in anything but the most arid stony soil or impossible clay pan if continuously well fed with inorganic fertiliser. A slow release fertiliser is probably unnecessarily expensive under these conditions. A really dry summer may produce casualties but there will be substantial rewards in most years.

Even in gardens where good conditions are available spare plants of perennial and rosettes species will often thrive in quite neglected herbaceous borders. I have often felt that a carpet bedding of *M. napaulensis* would be a marvellously innovative scheme for a local authority park. This could even be a lovely evergreen underplanting of a rose bed and the dedicated rosarian could always hoe them all out the spring they are big enough to flower. In this way, the stately flowering stems of the meconopsis would not compromise the show of roses; and if the bed is well-fed enough inorganically, both should flourish. Meconopsis of some species are easier and tougher than many imagine and should be experimented with. They are so easy from seed that no one should be deterred from trying them.

Difficult climates

It might be sensible of course not to try, but many gardeners, especially those who aspire to be plantsmen, wish to grow the rare and difficult as well as the commonplace. The fabled reputation of the blue poppy is very much an inducement to try. Growing meconopsis in a hot dry climate is no more unreasonable than my own desire to flower calochortus or oncocyclus iris.

The biggest problem for me, living in the United Kingdom, is knowing what constitutes a difficult climate. I have spent two years in southern Australia including a spell when it was over 45.5°C (114°F), I have seen the dry lands of the United States and Canada in summer and the beauty of New England in spring. A major problem is air humidity. A rich organic soil structure can be created and very often there can be a good supply of water from irrigation streams. In such climes this water is more often directed towards keeping the lawns green than keeping poppies blue. It should be possible, even in very dry summer conditions, to keep seedlings humid and shaded over the first summer in a frame. I suppose the species I would try first in such areas is *M. latifolia*. It might become prematurely dormant in summer but it should stay alive and achieve another burst of growth when planted out the following spring, even if it took three years to flower. *M. horridula* would probably be no more difficult and these two species should not be despised, both at best can be astonishing and *M. latifolia* is probably my favourite meconopsis. It may even be that *M. aculeata* can be adapted with a little trickery to growing with rude luxuriance. This species comes from the hot dry end of the Himalayas and may sometimes be subject to high daily air temperatures and radiant heat from the surface. This species has a reputation for difficulty, largely based on its need to keep its roots in damp conditions. Provided it has its roots kept wet in a good rich soil, then it may well take any amount of heat and dry air. It is increasingly being brought back from the wild and there are even reports that it grows in drier areas away from streams.

M. betonicifolia is tough and easily grown from seed so plenty of material is available. It is particularly subject to mildew in dry spells and, as with all fungal infections, there is little cure once well established. It is well worth while in dry areas spraying from spring with a systemic fungicide (but beware of scorch) from the moment they emerge. It should be possible with this species to keep it growing in the first year in an enclosed and humid frame and then hope to keep plants going for long enough in the second year to establish a perennial plant and keeping it alive over summer with ground water. The rosette-forming species are again easy to grow from plentiful seed and could be kept over summer in a humid frame. These species are much less prone to summer wilting than *M. betonicifolia* and possibly they will thrive with good feeding and ground water. They are so easy to raise that they are certainly worth trying. There are suitable localities even in potentially hostile environments. The Dandenong ranges in Victoria can be subject to 38°C (100°F) and dry heat, and yet the multilayer canopy of *Eucalyptus* sp. and other native trees produce an almost rain forest habitat with both tree and epiphytic ferns. Admittedly there can be heavy summer rains but meconopsis of several species are grown there in gardens.

The other factor to be considered is low winter temperatures often associated with the continental type of hot dry summers. It has been suggested that meconopsis will not tolerate hardiness zone 6 in North America but this seems unduly timid an expectation. *Meconopsis* species are either low growing or deciduous and even if snow cover is not reliable they can easily be covered with a dry frost protection probably of a cloche and dried plant material. They will certainly survive –20°C (–4°F) in Scotland with the frost freezing the ground hard for weeks on end. It is not to say that there are no losses under such conditions but most of these could be obviated if a little thought and trouble was taken in the early autumn to reduce the risks. I guess one might have trouble with meconopsis in the tropics however!

Species enjoying scree conditions

The most obvious species for such conditions are the relatives of *M. horridula* but there are a number of other species, which, though quite happy in the rich peat bed in summer, need quick autumn ripening and drying out in winter. This environment would also be the safest place to start new species unless there was unequivocal evidence that they were woodland glade species. Screes often conjure up the idea of a stony slope with little or no organic matter. I have the gravest doubt whether this is of the slightest value to any type of alpine plant in garden situations. In the wild this habitat is common but, in all cases, there is likely to be a continual supply of underground mineral rich water. Such habitats are admittedly likely to be extremely poor in organic matter but nutrients including nitrogen may be continually supplied. It is very difficult to imitate this dilute underground feeding though I have no doubt that the trickle feed systems used for, say, growing tomatoes with an automatic dilutor could be adapted. No doubt one day I will get around to building a scree over a plastic base that allows the use of such technology.

The far easier alternative is the rich scree which is probably acid. The basic materials for this are broken porous rocks such as sandstone and a sedge peat. It is valuable to incorporate as much leaf soil as is possible as well. Coarse gravel can be used as well as, or instead of, broken rock. It is best if the surface is top dressed with large porous pieces of rock and it requires some skill and care in the choice of materials to make this look attractive. Ideally this scree should have north and south faces so that a whole range of planting exposures can be achieved. The more southerly the longitude of the garden the more northerly exposure of the scree. A certain amount of shade may not matter and, if a real freedom of choice is available, the midday sun should be avoided. It is as well to build into this scree mist nozzles so that humidity and watering can be controlled. The upper 15 cm (6 in) of the scree should not contain less than 50 per cent organic matter and there can be pockets with as high as 80 per cent. It is advisable to build the scree as a series of flattish areas separated by strata of large rocks rather than a continuous slope. The slope on both faces should never be more than 30°. It is as well not to make these screes too large or else to incorporate a really firm access path making all places accessible—leaping about from rock to rock is for goats!

The value of varied exposure is that tricky plants can be tucked into a whole range of micro-habitats offering different growing conditions within as small an area as a square metre or so (a few square feet). I cannot emphasise how important imagination is in preparing these beds. A single rock that can be lifted in one hand will change the temperature-holding properties, alter wind speed and direction, produce differential effects of rain and watering and provide a cline of soil conditions from a cool root run under the rock to a baked soil a few centimetres away. One only has to look in the wild in any area where rocks are exposed to see how significant changes in micro-climate occur over a very small area. An imaginative scree can present to a dedicated plantsman in a few square metres all the pleasure that only seems possible with the yearned for hectares of woodland glade.

The one problem with this scree is feeding. If the climate is naturally of high rainfall then nutrients will be quickly leached, and in dry climates the need to keep the bed regularly misted during the growing season creates the same effect. It is not possible to top dress because it destroys the structure and anyway so many plants are adapted to grow flat on such screes. These slopes are also much loved by birds such as the blackbird and the stony top surface is a good deterrent to their digging habits. The main reasons for a strata of flat beds is the bird problem, since anything on a slope rapidly ends up at the bottom. In dry weather the misting of such screes seems to act as a magnet for birds from far and wide and in spring it may even be necessary to net the screes. The most sensible way to apply food is with a resin-coated slow release fertiliser in early spring and again in late summer. This is expensive but unless the scree is very large, little is required. This works extremely well and produces excellent growth in meconopsis and robust plants that survive the winter well.

In the wild these screes are likely to be either snow covered in winter or frozen dry. In a really wet climate with much winter rain even the fiercest drainage underneath the rich scree will leave the top soggy. This is of no consequence to forms of *M. horridula*, *M. latifolia* or *M. quintuplinervia* but for species like *M. sherriffii*, *M. punicea* and *M. discigera* and probably *M. simplicifolia* the surface must be kept dry in winter. It is not easy to place small tent cloches on a stony scree and the best course is to place a transparent roof at least 1 m (3 ft) above the scree or at least over that part with the choicest plants in. The solution is to have four clay, plastic or metal pipes 30 cm (1 ft) or so deep and at least 7.5 cm (3 in) in diameter, buried in a rectangular shape within the scree and disguised over summer with strategically placed pieces of rock. In autumn four uprights of wood are dropped into these holes and a sheet of corrugated clear plastic screwed on to the top held by a frame. It is either necessary to provide a gutter or arrange the scree such that the drips all fall on a plant-less rocky soak away (Figure 5). This may seem a lot of fuss but for the difficult plants it is essential and many other choice species of different genera also love this treatment. It allows drizzle and snow to blow under, and perfect aeration, yet it deters winter birds from digging. There

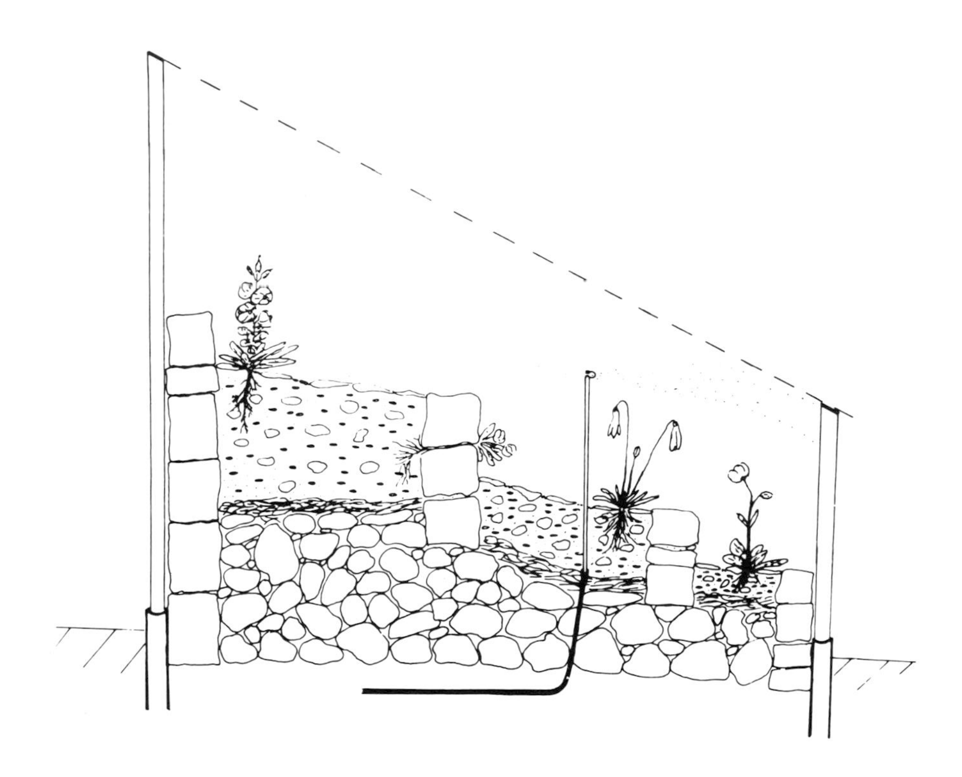

5. Misted scree

will be a cline of dryness from wet outside to really quite arid in the middle. This can of course be exploited once one realises the individual needs of different types of plants but it must be recognised with a cover occupying a couple of square metres that the middle can dry out completely and needs misting even in winter. It is possible to tuck bundles of dry bracken or even wide mesh sacks filled with expanded plastic chips over dormant plants as frost protection. This is often necessary with *M. sherriffii.* Dryness allows many plants to survive frost much better and some petiolarid primulas stay in flower in this bed even with 15° of frost. The biggest single problem, and perhaps the over riding one with the whole of this genus, is how and when to ripen plants in autumn. This is usually not difficult with seedlings in the first year except for the monocarpic form of *M. simplicifolia*. It is however critical with the difficult and potentially perennial *M. sherriffii* and with the monocarpic *M. discigera* (and one guesses its relative *M. torquata*). Time may prove that *M. punicea* is also an autumn awkward species. The simple answer is cover sooner rather than later but it may vary from year to year. Naturally drying winds and autumn sunshine are marvellous and should not be denied to them and covers should go on late in such weather, but remember one soggy day negates all your effort. The best advice is to place these species in the most free draining soil and the place most exposed to the drying influence of the prevailing autumn winds.

It is a very satisfying way of growing choice high altitude Himalayan plants, allowing many treasures in a small area, and the rewards are worth all the effort. I sometimes wonder at the attitudes of people who look at these contrived habitats or even a fan-trained fruit tree. 'It must take a lot of time they say.' 'Yes', one answers, 'perhaps half an hour a year for the alpine screes and less time to prune the cherries than to eat them.' The problem is not time but organising oneself to do these jobs and there is nothing like a gardening diary to help one.

Screes in difficult climates

This technique can probably be adapted anywhere and a totally artificial scree erected to prevent dry desiccating winds burning excessively; in the same way artificial shade could protect from too prolonged sunshine. Providing modest amounts of piped water were available the misting could even be under automatic control. It is likely however even in very dry continental summer areas that *M. horridula* (except the highest altitude forms), *M. latifolia* and *M. aculeata* would grow as long as the roots were kept wet and *M. quintuplinervia* might too in a shady north exposure. A transparent cover and dried plant material such as bracken or the artificial snow of expanded plastic chips method underneath it would allow the *M. horridula* types to survive really very severe low temperatures in winters probably even in hardiness zone 5 in North America.

SUMMARY. In the two basic habitats of rich peat bed or rich scree will grow almost all the species that we have, and would be the basis for adapting for species that may one day be introduced from the wild. Many species have their own particular foibles and some are certainly more difficult than others. These basic techniques should therefore be modified as necessary, as discussed under particular species in Chapter 5.

3 *Cultivation*

Seed collection and storage

A few *Meconopsis* species may be propagated vegetatively (see under the relevant entries in Chapter 5), but in most cases the propagation of meconopsis will start from seed. There are various sources of seed available, but only a few species are readily obtainable. The most obvious source is from one's own garden and, as many of the species are not perennial, constant renewal from seed is not only desirable but essential. The seed of some species is regularly available from major seed houses: this is likely to include *M. betonicifolia* and probably *M. napaulensis*. There may at times be other species such as *M. superba*, but the appearance of these scarcer species probably depends on good amateur growers harvesting a crop and passing it on to the seedsmen for sale. Seed collected from the wild is also available commercially from specialist seed sources which may be based in India and use native collectors.

The best selection of seed by far comes from the seed exchanges of various gardening clubs (see Appendix). In Britain these are the Scottish Rock Garden Club and the Alpine Garden Society: they may offer 15 or more species and varieties including species that are very rare. They have offered more than 20 species in the last decade. The American Rock Garden Club is another source of seed. All these schemes will provide seed for members in any country around the world, subject only to any import restrictions on the particular species. The schemes are run entirely by amateurs and, considering the number of different varieties can run to 4,000 or 5,000, the effort required is enormous. Membership is necessary to take part in these schemes and while many people obviously donate seed it is not essential. Seed is distributed according to various guidelines but donors usually have priority. Seed of many of the more commonly cultivated species of meconopsis is usually produced in abundance and more seed of *M. betonicifolia* may arrive, several kilograms weight, than of any other specimen of rock garden plant. The seed is distributed very early in spring and allows for timely sowing. The main problem with such seed is that the identification of the species yielding seed is dependent on the ability of the donor to assign it correctly. It has to be confessed that much seed is incorrectly named and the whole problem is compounded when people grow such seed on and re-submit it for distribution as the incorrect species.

The last way in which seed may be obtained and by far the most exciting is from the wild. People are increasingly travelling to various parts of the Himalayas and China and many collect seed on a casual basis or in some cases as part of fairly serious botanical expedition. Seed has recently been brought back from the western end of the Himalayas, Nepal, Sikkim, Bhutan and parts of China. At present the triangle between north-east India, Burma, Tibet and China is the least accessible and is perhaps the richest area. There have nevertheless been some quite superb re-introductions in the last few years of desirable species and forms. There is the prospect, at least from China, of obtaining completely new species. Some of this seed finds its way into the seed exchanges when first collected and will almost certainly be widely circulated if it is grown on successfully. In many ways amateur plantsmen are significant in establishing new and difficult plants, and once growers develop a reputation for particular types of plants they are often supplied with seed by the professional botanical gardens.

It is regrettable that the storage of some meconopsis seed is not straightforward and loss of germination potential is the consequence. The Kew seed bank feel that a major problem with meconopsis seed is the immediate post-storage conditions but if these are right then viability can be maintained in store for years. The viability of even common species drops radically after the first year unless the storage conditions are precisely maintained. The seed should be kept desiccated dry and cold and details of the conditions for storing and subsequently germinating stored seed has been described by Thompson (1968), including the use of growth hormones to break dormancy.

The amateur should in general collect seed and dry it prior to storage in a cool and dry room. All the chaff should be removed as soon as possible since this chaff will contain spores of disease organisms, particularly fungi, and this will contaminate the seed. Much of the potential for seed contamination will be carried over however careful the seed collecting process but there is no point in practising anything less than the most hygienic conditions. The seed of many species of meconopsis is very similar and careful labelling should be maintained at all stages (Figure 6). Seed of this genera is heavy when viable and winnowing over a large sheet of paper on the kitchen table is a quick way of doing the job. A flour sieve only allows a proportion of smaller seed through and almost none of the large-seeded species. A series of graded sieves would be helpful but they are not usually available outside professional circles.

Seed cleaning is a messy business especially with meconopsis that have masses of spines. Winnowing fills the air with dried spines which irritate skin and eyes and are probably bad for one if inhaled. It is possible with a lot of seed to place the unsorted seed on a sheet of stiff paper inclined at a slight angle; a gentle tap, and the heavy viable seed will run to one edge leaving the chaff behind. If masses of seed is available then one does not need to worry if some viable seed is thrown away with the chaff but for rare seed one just has to draw on one's daily supply of patience.

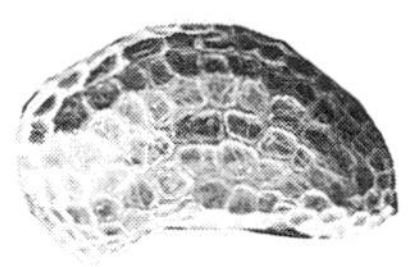

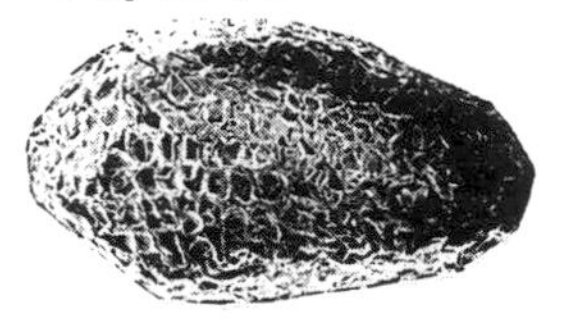

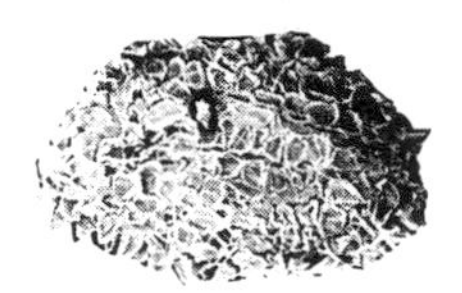

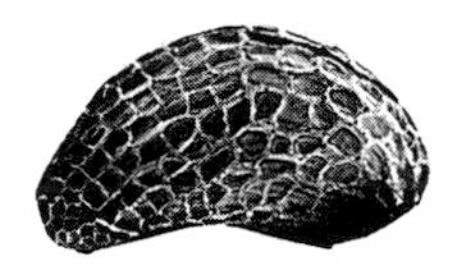

6. Scanning electron micrographs of various seeds to scale (note for example that *M. betonicifolia* is distinct from *M. grandis*)

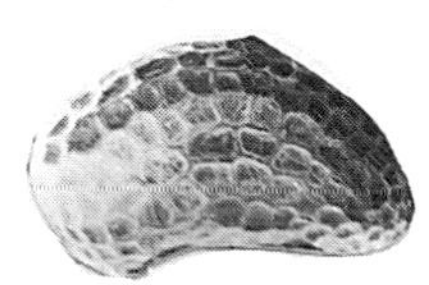
Cambrica

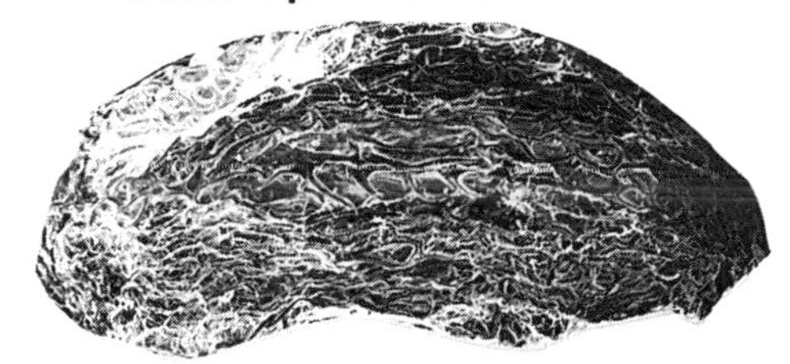
Quintuplinervia

Betonicifolia

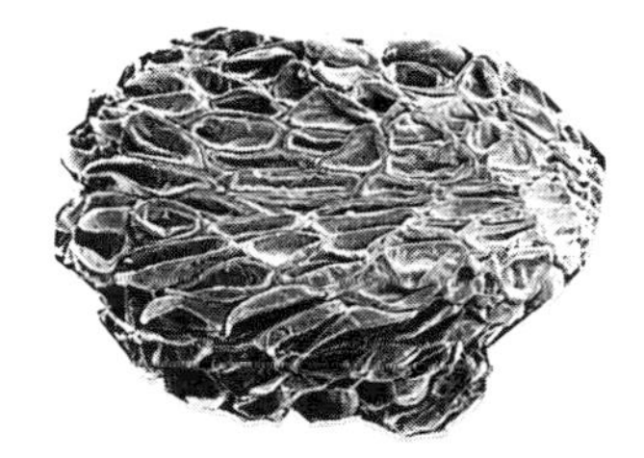
Punicea

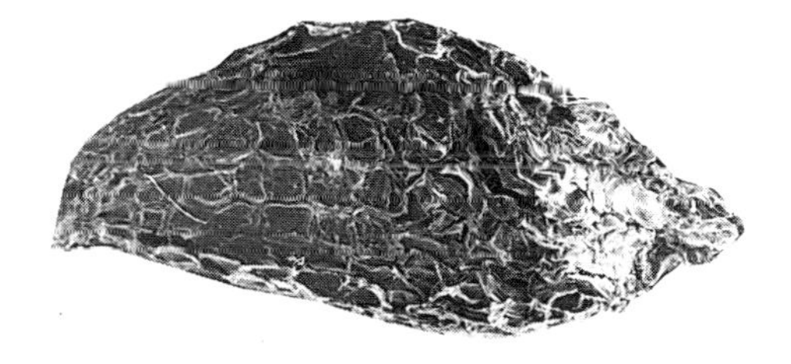
Grandis

Integrifolia

x Sherriffii

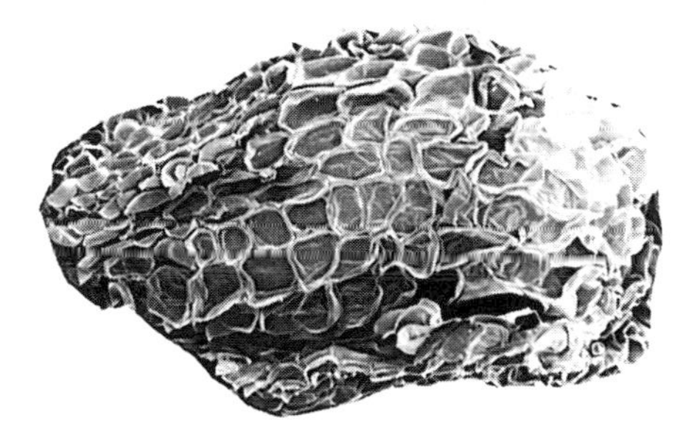
x Sheldonii

Figure 6B

Some species produce seed that is difficult to keep in a viable condition. This is a phenomenon for many species of Himalayan plants and is particularly associated with primulas. Needless to say it is always the most desirable

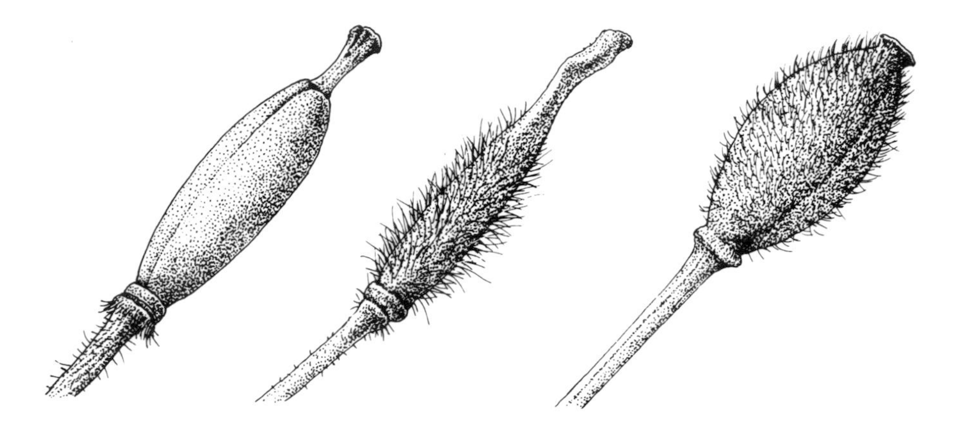

7. Seed pods of *M. grandis* (left), *M.* × *sheldonii* (centre) and *M. betonicifolia* (right)

ones! The reasons are not easy to assess, though the problem has long been known. It is strange since many of the plants involved are not long-lived even if theoretically perennial. The rapid loss of viability cannot easily be seen as an advantage. The most likely explanation may be that collecting and drying seed is so alien to the natural processes that a deep and perhaps irreversible dormancy is induced in collected seed which does not occur in the wild. It might also be that germination of fresh seed in the wild is normally so successful, and the monsoon climate so regular, that the need for longevity in the seed has never been the subject of selective pressure and has simply not evolved. Kingdon-Ward suggested that the season was so short for high altitude species that they did not have time to produce food reserves, but the phenomenon is most apparent in some of the large-seeded Chinese meadow species. It is also possible that these seeds have two strategies, one of immediate germination before drying and then a long-term dormancy, lasting perhaps years, that is broken by factors which we do not understand. So much effort is required to collect seed from the wild, that we need an answer to the problem, an answer that will only be forthcoming with serious research into dormancy strategies of the difficult species. No effort should be spared to achieve the conservation value of a genetic reserve of material which can be stored almost indefinitely.

In the short term rather more imaginative ways should be devised for bringing seed home. This is of course more easily said than done, since any equipment has probably to be carried over long distances by personal human endeavour. Thoughts on this problem are not new and it occurred to plant collectors early in the last century who were already well aware of poor results from collected seed; some of that sent back by Forrest had little success. Recently, the Alpine Garden Society Sikkim expedition brought back seed of *M. discigera* for me that was undried and stored in damp sphagnum moss with a fungicide. This grew on excellently when sown on arrival, so at least the technique did not kill the seed but one has to say that dried seed of this species did just as well. There is no doubt that the first approach should be to devise methods of bringing back the seed 'green' of species that are difficult to re-awake from dormancy.

In practical terms this means that the seeds of certain species and any new seed brought in from the wild should be sown, in part at least, as soon as possible after collection. It appears essential for the two closely related species, *M. quintuplinervia* and *M. punicea*, and probably for *M. simplicifolia* in some forms. This seed will not in general germinate the autumn it is sown but come through when placed in heat with the rest of the species early the following spring. The seed should be sown in the recommended composts (see p. 35) and lightly gritted to prevent the growth of liverworts over winter. The pans should be kept just moist in a protected frame. Frost and snow will not hurt them and may even be beneficial, but leaving them to become sodden by long periods of winter wet would most certainly be detrimental. Recent evidence with *M. punicea* shows that with spring-sown seeds failure may be total but there is some prospect of germination a year later. It is thus important with new, rare or difficult materials that the seed pans are kept for at least a year. This requires that the surface is kept free from other growth and may mean that routine hand removal of it is required. This is a most fiddly and time-consuming job but compared to the hours and efforts involved in collecting new seed from the wild we owe it to the collector.

There are one or two other circumstances for sowing in the season the seed is harvested. In the case of seed collected in the garden from perennial species, particularly *M. grandis* and *M. betonicifolia*, if it is sown about three weeks after harvest in late July, germination can occur rapidly and the plants be grown on to a reasonable size of perhaps 5 cm (2 in) in height before the onset of cooler autumn weather. These should then be overwintered in a frame with good air circulation and kept just moist. They would be best left in a pan having been pricked on as soon as possible after germination. They will become totally dormant below ground level in winter but most will re-emerge in spring with the increase in temperature. They should then be potted on and kept growing. The aim of this is to produce really robust plants by the end of this second year that will flower the following year as strong plants without undue risk of their dying after flowering. It is possible that this technique is worth trying for the same reason with rarer species like *M. sherriffii* and some forms of *M. simplicifolia* but the risk of overwintering the seedlings might outweigh the gains of stronger plants.

The second circumstance for autumn sowing is of wild-collected seed from the previous year. Many more people are now able to take flower holidays in remote parts of the world and will normally travel at the best time for seeing plants in flower. They do occasionally find capsules of seed still intact from the previous year. This seed is already approaching a year old and it is probably sensible to plant at least part of it as soon as possible. This may well then germinate and require to be overwintered. Recent seed in this category from China of a rare species was sown and germinated but had gone little beyond the emergence of two tiny true leaves from the seed leaves. These minute seedlings then became dormant but the eye of faith (or a good microscope!) showed a literal pinpoint of green in the middle. The roots were probably less than 1 cm (½ in) into the compost at this stage and great care was required to keep the compost just moist. It is certainly essential to keep severely dehydrating and cold winds from such seedlings. Regrowth in spring of some of these was rapid and good transplantable seedlings were in evidence by early April without heat. The type of expedition which collected herbarium material in flower and then returned to collect seed from marked populations will in general be too expensive to contemplate. The Alpine Gárden Society expedition to Sikkim did in fact make two trips in 1983 but even these valiant efforts were to a certain extent compromised by the terrible monsoon conditions that extended late into that autumn.

In general, collection of modest amounts of wild seed is not likely to be harmful but substantial collection of rare species especially if the species is highly localised would almost certainly be detrimental.

SUMMARY. Meconopsis seed is not long-lived unless stored under controlled conditions. It is safe to store seed from one summer to the next spring in dry and cool conditions. *M. quintuplinervia*, *M. punicea* and some forms of *M. simplicifolia* should be sown the autumn after harvest as viability is rapidly lost. Some seed of new species and re-introductions from the wild should also be autumn-sown. Collection of seed in the garden should only be made from ripe capsules of good forms of each species.

Seed sowing

Nearly half the species in cultivation will quite happily seed themselves in the garden but for a number of reasons it is better to harvest and sow seed of most species. There are great advantages to using a heated frame in early spring. It is quite possible to sow seed of some species in a simple seed sowing compost but, generally, quicker and more reliable germination is obtained in a different type of medium and for some difficult species it is virtually essential. Mike Hirst, in charge of the national collection of *Meconopsis* at Durham Agricultural College, ran extensive trials using seed of *M. betonicifolia alba* with a whole range of growing conditions. The most important finding was that a peat-based soil-less compost produced a much quicker germination. There is an account by Thompson from Kew (1968) on meconopsis germination with much valuable information relating to temperatures, light and the use of plant hormones as dormancy breakers. I have also carried out much amateurish research myself and the general programme outlined below is based on or confirmed by quite a substantial body of knowledge.

The compost should contain much expanded organic matter both to retain moisture and hold water. A proprietary peat-based soil-less seed compost is satisfactory for most species, providing the very coarse lumps of fibre that the material inexplicably sometimes contains is sieved out. It helps if a good coarse grit is incorporated and if the amount of this is increased below the surface 1 cm (½ in), so much the better. It is also useful to incorporate one of the expanded rock materials such as Perlite which increases aeration and drainage properties at the same time. There is no doubt that some really good leafmould can have especially beneficial effects when the seed has germinated but there is substantial risk of introducing various pathogens and this is definitely undesirable. If the leafmould can be sterilised at a temperature that is hot enough to kill pathogens but not so hot as to break down the structure then this is probably acceptable. A significant additive which is necessary for some difficult species is dried sieved sphagnum moss. One is always advised to obtain the deep red form from open sphagnum bogs, but few of us have access to such places and have to be content with small quantities of whatever we can find. This material is best sun-dried to complete crispness and then sieved through a coarse sieve. The great virtue of this material is that it has great water-holding capacity as well as trapping much air. The sown seed nestles in between the fibres of the compost and a fully saturated humidity is thus maintained at all times in the spaces in the compost. Total saturation of the atmosphere is very important with many species when they are in active growth and for some seedlings it is critical. A moss peat compost is good in having these properties too but sphagnum is undeniably much better. The sphagnum in the compost in general does not become too rapidly taken over by moss and liverworts though it will eventually happen.

The recommended mix would be 50 per cent soil-less compost or plain sieved moss peat (sedge peat forms a hopeless skin on composts and should be avoided), 30 per cent sphagnum and 20 per cent coarse grit or expanded rock granules. The sphagnum can be left out at depth in the compost if it is in short supply and extra peat or soil-less compost added. It would be expected that a proprietary soil-less seed compost would contain traces of added inorganic food and this is undoubtedly beneficial. If a seed compost is being made up from plain peat and the other ingredients described, then small amounts of such as John Innes base might be beneficial but no more than a teaspoon to a 4½ litre (1 gallon) bucket. The alternative would be to use a quarter strength plant food watered on to the pots once germination has occurred. It is essential that the compost remains moist at all times, drying out once germination has occurred is fatal. Very often most seed germinates at the same time and losses can be total. The advantage of sphagnum in the compost is that drying out is a slow process, with water being drawn from below, while the micro-climate around the seed dries out very slowly. There is little doubt that meconopsis seed will rot in saturated conditions and overwatering should be avoided too. Growing the seed in the above compost in mist conditions with soil warming cables would be ideal. The misting needs to be adjusted so that the pots are kept just wet and not saturated.

All species of meconopsis require some light for germination. They must be sown thinly in the compost as most species readily germinate and overcrowding is a certain cause of trouble. Small square pans are easiest to fit into a frame but 7 or 10 cm (3–4 in) round plastic pots are quite satisfactory. If viability of seed is doubted and a good deal of seed is available of a precious variety then the seed should still be sown thinly but over the surface of a much larger pan. Small quantities of rare seed should be sown individually in small pans with exact seed spacing so that the most rapid and healthy growth can be obtained (about 50 seeds in a 7 × 7 cm (3 × 3 in) pot). The seed pans need covering with a faint dusting of very fine compost, or for autumn-sown pans for overwintering, a single layer of fine grit not much larger than sand. Much of the seed in an open sphagnum compost drops into little crevices when sown but compression does take place with watering and seed can become exposed. Large seed of species such as *M. grandis* can be covered to a depth of 2–3 mm (⅛ in or so) but finer seed such as *M. bella* need the merest dusting.

Great care should be taken in watering. In the first place the compost is best soaked by immersing the seed pans to half their depth in a water container and the surface misted with a fine spray. A rose on an ordinary

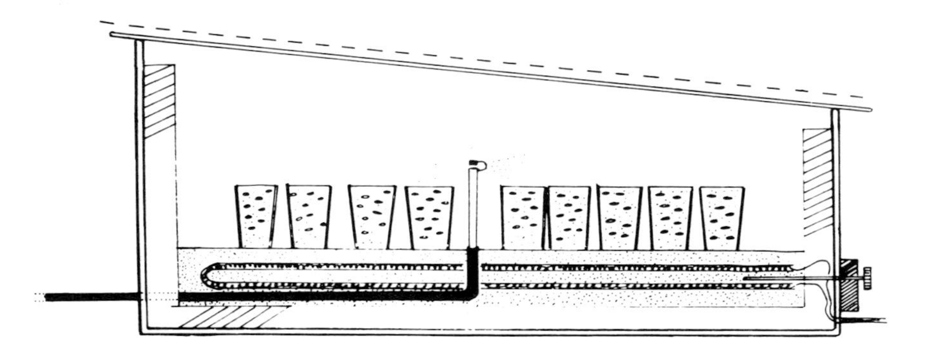

8. Heated and misted seed frame

watering can produces too coarse a spray and will wash all the carefully sown seed into corners of the pots, or worse may even wash seed into the next pan leading to considerable confusion. Watering should be undertaken with great care until the cotyledon leaves are fully expanded and the plants well-anchored by the emerging roots. A mist nozzle buried in the frame is ideal and a simple home-made system is illustrated in Figure 8.

It is not essential to use heat but there is little doubt that it is most advantageous. Sowing should be carried out in mid to late February. This is often the time of year that seed is distributed from the seed exchanges. If heat is not to be used then sowing should be delayed for a month to six weeks. A simple frame can be heated with a soil warming cable but a thermostat is advisable. This should produce a minimum of 8–10°C (46–50°F) by night and a maximum of about 20°C (68°F) by day. The temperature should not be allowed to rise much above this even for brief periods and a sunny spell will require the frames to be shaded and perhaps opened. Great care must be taken never to let the seed pans dry out and a sunny day with cold drying winds is especially risky. Germination of most species will take place rapidly within two or three weeks and if seed has not germinated after that period it probably will not germinate at all. The seed leaves of most species are very alike and after a time become recognisable from nearly all other genera. They are about three times as long as broad with a rounded end and are held horizontal. The seedlings of *M. integrifolia* have very large leaves and only *M. bella* is different in that the leaves are proportionally much narrower but they are still distinctive. The seed leaves of *M. horridula* and relatives are rather more glaucous and of a very slightly bluer shade of green.

Some species are prone to damping off and some precautions can be taken against this. If a soil-less compost and dried sphagnum is used it should be sterile and sphagnum has, at least by reputation, some antiseptic properties. It is only if leafmould has been introduced that pathogenic organisms may be present in the compost and, as already stated, careful sterilisation of this is sensible. There is no doubt that however sterile the compost, debris in with the seed and even the seed coat itself may be contaminated. It is possible to surface-sterilise seed with a dilute chemical solution, but this should only be undertaken if careful instructions can be followed. I have no reason to suppose that a proprietary seed dressing shaken in among the seed before sowing is not valuable. There are systemic fungicides, such as benomyl, and colloidal copper solutions that are effective on seedlings and some are specially formulated to counter-act damping off but great care is needed in using these since meconopsis seedlings are unduly sensitive to many chemicals. It is obviously better to try chemical warfare against pathogens rather than lose the whole batch of seed but use cautiously dilute applications to start with.

SUMMARY. Clean seed should be sown in a well-drained and aerated peat-based soil-less compost preferably with dried sieved sphagnum and extra grit added. Seed should be started in early spring with modest bottom heat not exceeding 20°C (68°F) and with moderate levels of light admitted. The seed should not be allowed to dry out at any time and seedlings protected from desiccating winds. Seedlings should normally grow on rapidly and show true seed leaves within a week or so. The most rapid growing species is *M. integrifolia* and the *M. horridula* group is slowest with the exception of *M. bella* which is even slower.

Growing on

The plants can be pricked on in six to eight weeks from germination into pans at least 7.5 cm (3 in) deep. It is best to shake the seedlings out of the seed pan and the roots normally fall free in the loose seed compost. Pricking on should be done when there are two strong true leaves. The stems are very soft and should not be touched, and only the edges of leaves lightly handled. Seedlings of *M. integrifolia* will be approaching 1 cm (½ in) but those of *M. horridula* will be much smaller. Seedlings should be spaced 5 cm (2 in) apart and the compost lightly tapped into the planting hole and the plant then firmed very lightly. They must not be planted into compressed compost or pressed firmly in. The compost at this stage can either be a proprietary soil-less compost or a home-made mixture of peat (7 parts), grit (3 parts) and leafmould (1 part) with added John Innes base at about the No. 1 level. It is necessary to incorporate some feeding in the compost. The spring weather at this time can be severe and it is best if they remain covered with shading from bright sunlight and shelter from winds. The seedlings handled at this early stage show little sign of check if treated gently, but *M. latifolia* and *M. aculeata* can be quite slow to grow on. Seedlings of all species are vulnerable to aphids but are sensitive to many chemical sprays and can be burnt by some systemic insecticides, especially if applied during a dry sunny spell. The answer is to spray with several very dilute doses of a systemic insecticide and carefully monitor the effects. I suspect the problem is that the insecticide lodges within the very water repellent hollows of leaves and becomes unduly concentrated and washing the insecticide off before this happens may be the answer.

The pricked on seedlings will make rapid root growth into the rich light compost and in a matter of weeks should either be planted out or potted. The rosette-forming and the perennial species make a large root ball and even the tap-rooted species that are winter dormant, such as *M. horridula*, grow many fine roots. The compost is thus bound together by the roots and once they fill the container each seedling can be scooped out with the root ball intact and placed in a 10 cm (4 in) plastic pot already half-filled with similar compost to that used for pricking on. They should not be packed down tightly but left to settle with watering. The pots can then be placed in a wind sheltered spot with shade for most of the day. The more humid the atmosphere the better the growth of the seedlings, but when a large number of pots are watered in the same area they tend to create their own humid micro-climate and growth is normally rapid. It is possible to plant out directly from the pricking off box providing two conditions can be met. The planting out site must remain humid at least until the plants are well established with roots deep into the soil. This can best be achieved using a mist unit in the bed morning and evening (very little actual water is required). The plants must also not be overgrown and shaded otherwise they will became lank and drawn. It is so easy in mid-May to see large places for planting in gardens only to find they disappear as summer progresses, and what were thought to be well behaved and refined plants nearby become territorially ambitious especially encouraged by the food and water so lovingly provided for the new meconopsis.

Plants potted on are of course much safer if a dry summer occurs or in areas where hot dry summers are predictable. An enclosed humid area could probably be created to save many dozens of potted plants, even when watering restrictions put many peat garden plants in jeopardy. They will generally have well filled 10 cm (4 in) pots by mid August and it is beneficial if a small quantity of slow release fertiliser has been added to each pot. This is the best time to plant them out since they will then rapidly establish themselves with a plunging root system before winter, and still have time to ripen for winter dormancy. If dry conditions still exist at this time they are large enough to cope as long as an occasional watering is given.

Meconopsis are never really happy in pots and however much they are potted on they will eventually suffer lack of space. Because of this I have advocated planting them out as soon as possible. Sometimes plants have been left overwinter for a variety of reasons and although the growth is stunted they do usually grow on quite satisfactority when planted out in late spring of the following year. The species that flower as biennials will then sometimes flower almost immediately and often rather disappointingly. This is not an invariable rule and a proportion of *M. horridula* and *M. latifolia* will grow on and produce flowering plants in their third year. There is certainly a risk of a high proportion of the potentially perennial species such as *M. betonicifolia* flowering immediately and almost certainly dying after flowering. The rosette species that are monocarpic however are the most tolerant and will stay alive and stunted for up to two years in quite small pots while their planted out siblings are flowering plants 2 m (6 ft) high. It is possible in areas where these rosette species are vulnerable to winter wet that overwintering them the first year in pots might even be advantageous. They do need care in winter to keep both the deciduous and the evergreen species moist and free from raw, cold, desiccating winds. I always keep a few in pots as replacements for the occasional winter losses.

SUMMARY. Seedlings need to be pricked into a rich, free-draining peat-based compost as soon as two good true leaves are formed. The compost must not become compacted and shelter from wind and excessive sun provided. Plants should be rapidly potted on into a similar mixture or planted out. Potted plants should go out into the final site by mid-summer. Some plants can be overwintered in pots but it is best avoided.

4 *Plant Associations*

Meconopsis are generally regarded as plants for the peat garden and they are to be found there in many of the famous gardens in Scotland. It is possible to grow them in other sites, even in the herbaceous border, but there is little doubt that they look well mixed with many classical peat garden plants. They are however not easy to accommodate in peat beds with other plants unless some very basic features of cultivation are understood.

Many peat garden plants are from woodland, and in this habitat there is a constant deposition of leaf litter from above and penetrating roots from the trees below. This leaves the narrow but immensely fertile layer on the forest floor with a hard pan underneath. Many plants have adapted to exploit this thin habitat and are characterised by fine spreading roots within the surface later only. Plants in this category include cypripediums, some primulas, iris species, choice ferns, many of the most desirable North American and Japanese woodland plants such as *Dicentra*, *Jeffersonia* and some trilliums. In some cases, such as the stem rooting lilies, the perennial part of the plant is deeper but annual roots exploit the surface layer. The great secret of cultivating these plants is an annual top dressing. This is preferably of good quality leafmould, but certainly of some well-decomposed organic matter. This dressing should not be peat itself since this is poor in nutrients. If a peat is all that is available then a well-decomposed peat should be spread as a thin layer with an inorganic fertiliser added. Pulverised bark is all very well as a sterile surface mulch but is worse than useless when any attempt is being made to grow small choice woodland species.

Other peat garden plants are more truly dependent on a peat environment and these are moorland plants, typically ericaceous and exemplified by many rhododendrons. These make massive fine fibrous root balls which exploit a large area of poor soil. It is still a matter of fact that many rhododendron collections are starved because the natural leaf litter is swept away to keep the place looking tidy.

There are a few genera such as erythroniums and some trilliums that penetrate deep into the soil of peat gardens and are much less dependent on surface mulching. In general, however, if the top 5 cm (2 in) of the peat bed is kept well nourished by the addition of well-composted organic matter you will grow many of the finest and most difficult plants exceptionally well. And in areas where the climate is marginal, the better the standard of cultivation the greater the likelihood of survival during adverse conditions.

These are regrettably exactly the sort of cultivation treatments that will not work for meconopsis. A really rich top dressing will help with monocarpic species that are relatively small but all the rest require lots of good organic matter incorporated deep into the soil. Every three or four years perennial species like *M. grandis* need to be dug up, divided and replanted into a well-dug and composted area. Places where the monocarpic species are grown need completely refurbishing. This is a totally different strategy to most other peat garden plants. Some, like the Himalayan primulas, also from lush alpine meadows, enjoy similar deep rich conditions.

Many woodland species detest being moved, trilliums and the North American cypripediums being prime examples. Lilies and particularly *Nomocharis* will often deteriorate because of root damage and, although ericaceous plants with fine fibrous rootballs will move quite well, it can often set them back and they can be difficult to re-establish in a dry spell. We need careful planning of the peat or woodland garden where particular areas can be set aside for meconopsis or primulas. The whole area can then be dug up every three or four years. It does not necessarily need separate beds, just accessible areas that only grow plants that are compatible with meconopsis and need to rebuild and revitalise regularly. There are a number of other plants that can be included. The robust *Dactylorhiza* species like *D. elata* and *D. maderiensis* need regularly splitting up and so will *Fritillaria camschatensis*. Autumn gentians can be used as ground cover as these enjoy being split up and replanted. A good plan either on paper or in one's head certainly helps when maintaining a peat garden, so that a proper rotation of the various groupings of primula and meconopsis can be achieved on a year in, year out basis.

There is little doubt that the most valuable plants to associate with the perennial species of *Meconopsis* are liliaceous kinds. Dwarf species can be planted in front and very tall growing ones behind; in both cases, though, it has to be possible to dig up the meconopsis and renew the soil without disturbing the lilies.

There are two ways to grow lilies and relatives. Buy them in as bulbs, and grow the easy kinds, mainly hybrid, and face the almost certainty of virus-infected stock, or grow them from seed. Delicate plants like *Nomocharis* will rapidly succumb to virus as will many species of lilies and you will be left with only the more robust hybrids. These are of course perfectly acceptable but the sources from which *Meconopsis* seeds are available often supply many delightful species of lilies and related genera such as *Nomocharis* and *Notholirion*.

Alec Duguid, who once ran Edrom Nurseries, described a splendid technique for growing lilies from seed which I find nearly fool-proof for any peat garden lily types. He is now retired to Ballater not too far from where the

Queen has her summer residence in Scotland and once, when I was taken to meet him, showed me the secret which is immensely simple. Dry out a good bulk of sphagnum moss over the summer in a hot part of the greenhouse or even the airing cupboard. Rub this through a coarse sieve and fill a 20 cm (8 in) pot to the brim and press it down. If like me you always over-egg the pudding (to use a North Country expression) then you can add some leafmould. Place the lily seed on top and add another 1 cm (½ in) of dried sphagnum and keep damp. When the seed has germinated water with a half strength tomato (or similar) fertiliser about every two weeks in summer until they go dormant.

Continue this for about two years (only one with vigorous species like *L. regale*) and then plant the whole potful into the flowering site and sit back and wait for the gasps of admiration. The pot does not collect nasty liverworts in any hurry nor dry out at depth unless really neglected. *L. pumilum*, *L. oxypetalum*, *L. nanum*, *L. mackliniae* and *L. formanosanum* var. *pricei* are dwarf for the front of the planting. *L. regale* and similar species, the martagon lilies, American species like *L. canadense* and *L. grayii* are all easy and excellent to flower behind the meconopsis. *L. szovitsianum* is very easy and flowers in two years from seed and is essential for its beautiful perfume. *L. speciosum*, *L. auratum* and their hybrids and varieties, which are expensive even from seed, are brilliant for later in the season but take longer to reach flowering size. Start at least four pots of lilies a year, every year, and you will always have something new to anticipate flowering and this covers your impatience in waiting for those that take many years, like *Cardiocrinum giganticum*. There is only one golden and unbreakable rule to this and that is all lilies and relatives (seedlings too) must be sprayed every two weeks with a systemic insecticide against greenfly during the whole of the growing season. If you mix a fungicide into the can as well, it is the counsel of perfection.

5 *A–Z Listing of Species*

Meconopsis are late spring-early summer flowers but there are blooms from the end of April to mid-August if species are carefully chosen. The first to bloom is usually *M. integrifolia* and in a warm spring and a sheltered spot this can be the last week of April. The middle of May will see *M. integrifolia* at its impressive best and the first blue buds of *M. horridula* will be bursting as well as the less common *M. aculeata*. The first pristine white flowers will open on *M. superba*, usually the first of the monocarpic rosette species to flower. Farrer's 'Harebell Poppy' (*M. quintuplinervia*) has a succession of blooms from late May and these appear in flushes right into the autumn. The first week of June sees the classic blue poppies coming into their own with *M. betonicifolia*, *M. simplicifolia* and *M. grandis* plus the wealth of cultivars and hybrids at their best. They remain perfect for about 2–3 weeks but there are individual blooms well into July. *M. latifolia* also blooms in early June usually at least a fortnight after *M. horridula* appears. *M. napaulensis* and *M. regia* hybrids as well as *M. paniculata* flower from the top down beginning in the first week of June. Good forms have 50 or more blooms open at once and, as there may be up to ten or more flowers on each panicle opening in succession, they are attractive for many weeks before they gradually have a tired look with more seed pods than flowers. The last fortnight in July leaves one longing for the autumn tidy up though, unless seed is required, the monocarpic species such as *M. horridula* and the rosette species like *M. napaulensis* can be pulled up. *M. betonicifolia* can have the flowering spikes cut to the ground. *M. punicea*, at least in the light of my limited experience, is peculiar. It has a mad desire to flower that seems quite unrelated to climatic suitability. They flower from early March to November and they appear to insist on flowering until they die. They are however at their most spectacular best with all the others in June. Flowering in this genus seems to be very predictable from year to year and may well respond to day length far more than temperature; while they may flower earlier in places to the south of Scotland, it is probably not much earlier.

In the following A to Z listing of Meconopsis species, under the relevant species, details are given of Royal Horticultural Society Awards of Merit (AM) and First Class Certificates (FCC).

See also Appendix IV, where the colour of some species is described in terms of the RHS Colour Chart.

Key to species likely to be in cultivation

1	Leaves evergreen as expanded winter rosette	2
1	Plant at least partly deciduous	10
2	Flowers yellow	3
2	Flowers not yellow	6
3	Purple spots at base of spines on leaves	*M. dhwojii*
3	Leaves without purple spots	4
4	Stigma pale mauve, leaves dissected	*M. paniculata*
4	Stigma deep purple, leaves entire	*M. regia*
4	Stigma green, leaf dissected	5
5	Leaf highly dissected, single flowers on stems	*M. robusta*
5	Leaf highly dissected, several flowers per stem	*M. gracilipes*
5	Leaf moderately dissected, several flowers on stem	*M. napaulensis*
6	Flowers white	7
6	Flowers red, pink or shades	8
6	Flowers blue	9
7	Stigma purple	*M. superba*
7	Stigma green	white form of *M. napaulensis*
8	Stigma green, leaves dissected	*M. napaulensis*
8	Stigma purple, leaves entire	red form of *M. regia*
9	Leaves very regularly lobed (see Figure 14)	*M. violaceae*
9	Leaves irregularly dissected (see Figure 20)	*M. napaulensis*
10	Plants with yellow flowers less than 5 cm (2 in) diam	11
10	Plants with yellow or other colour flowers more than 5 cm (2 in) diam	13
11	Base of plant without hairs	*M. cambrica*
11	Base of plant with rufous hairs	12
12	Flowering stem unbranched	*M. chelonidifolia*
12	Flowering stem branched	*M. villosa*
13	Very small plant with glaucous divided leaves	*M. bella*
13	Larger plant with hairy or spiny leaves	14
14	Leaves with stiff bristles or spines	15
14	Leaves with soft bristles or hairs	18
15	Long strap-shaped leaves	16
15	Broad simple or dissected leaves	17
16	End of leaf with distinctive notch (Figure 14)	*M. discigera*
16	End of leaf simply rounded	*M. horridula*
17	Leaf not dissected	*M. latifolia*
17	Leaf lobed or dissected	*M. aculeata*
18	Yellow flowers	*M. integrifolia*
18	Cream flowers	Hybrids—*M. × beamishii, × harleyana, × sarsonsii*
18	Pink flowers	*M. sherriffii*
18	Red flowers	*M. punicea*
18	Blue or purple (or white) flowers	19
19	Stamens of flowers blue	*M. simplicifolia*

19	Stamens of flowers green	20
20	Flowers from basal scapes	21
20	Flowers not from basal scapes (or if so large, 7.5 cm (3 in) +)	22
21	Flowers mauve-lilac	*M. quintuplinervia*
21	Flowers muddy red-purple	*M.* × *cookei*
21	Flowers cream	*M.* × *finlayorum*
22	Flowers with characteristics as Figure 9	*M. betonicifolia*
22	Flowers with characteristics as Figure 9	*M. grandis*
22	Plant keys out to *M. grandis*, small empty seed pod covered in rufous hairs	*M.* × *sheldonii*

NOTE *M. paniculata*, *M. regia*, *M. napaulensis*, *M. gracilipes* and *M. robusta* are all very similar except in detail of the leaf and they hybridise. Absolute certainty with yellow-flowered rosette species needs experience.

9. *M. grandis* (left) and *M. betonicifolia* (right), showing differences between leaves, flowering scapes and seed pods which together are reliably diagnostic

M. ACULEATA

Blue-flowered deciduous monocarpic species in cultivation (Plate 1).

This is a western Himalayan species that is monocarpic and winter dormant. It is closely related to *M. horridula* and *M. latifolia* but appears to be more difficult than either. It has lobed and dissected leaves of a blue-grey colour. The degree and shape of the lobing in *M. aculeata* is variable but in all cases it is distinctive. The leaves are sparsely covered with pale straw-coloured hairs and are quite prickly to the touch. The root system is normally a single major tap root but in very rich cultivation conditions this may become modified to a substantial fibrous rootstock. The flowering stem is 60 cm (2 ft) or more tall and covered in spines. The flowers of those at present in cultivation are a purple-blue, although sky-blue have several times been reported from the wild and were once in cultivation. Often in the genus

10. Comparison of leaves of *M. latifolia* (left), *M. horridula* (centre) and *M. aculeata* (right)

Meconopsis purple-blue is made to seem unattractive but in this species it has all the iridescence of shot silk and is especially beautiful. The flowers are usually of good size up to 6 cm (2½ in) in diameter.

There is a rather difficult taxonomic problem with this species relating to the colour of the stigma. This may seem a characteristic of botanical interest only but it is a valuable and readily visible one in the diagnosis of a number of species. Taylor does not ascribe a colour, so one must assume it is green. An illustration from the wild describes some plants with purple stigmas. In cultivation both colours are present as well. *M. horridula* has a green stigma and *M. latifolia* a deep purple one. I suppose the logical conclusion from all this is that it is a variable characteristic in *M. aculeata*.

Until recently it seemed likely that many plants of *M. aculeata* were hybrid, probably with *M. horridula*. This manifested itself as a very variable degree of dissection of the leaves tending towards the simple strap shape of *M. horridula* and very little fertile seed. There have been a number of new collections of wild seed from the NW Himalayas and it is now well re-established.

In various literature sources, *M. aculeata* is described as a plant of stream-sides or boggy ground. This is significant in a plant that comes from the drier end of the Himalayas, and may well account for the relative difficulty of the species in cultivation. It must be remembered that even in these dry regions there may be percolation of winter snow melt underground. This plant ideally requires a dry-surfaced scree with water percolating underneath in summer. In practice it grows well in a good damp spot as long as this does not become waterlogged in winter. It is possible to create this artificially by growing them in a 30 cm (12 in) deep container with good drainage material at the bottom and a plug hole. The plug is taken out in winter and the tub covered. This may seem a bit contrived but all the mechanics can be landscaped so as to appear invisible and it only requires a few minutes attention to the plug hole twice annually.

This technique is actually quite useful for a range of Himalayan plants, such as some primulas, which spend the summer in boggy conditions but need to be winter dry. Whatever the solution to keeping them damp in summer, it is likely to lead to leaching of nutrients from the growing medium. There are two ways of replacing nutrients: first, as the plants are usually biennial the bed can be remade with new organic material after each generation is finished. The alternative is to use a good resin-based slow release fertiliser during the summer months applied as growth starts in spring. The advantage of this species in many climates is that it is tolerant of a very dry atmosphere during the growing season and is not subject to mildew.

There is no suggestion that this species is not frost hardy but a mild early spell can bring out small underground slugs which devour the growing points just as they emerge from the dormant root stock. A really sharp and gritty top dressing reduces the threat from these pests. This species has a touch of class compared to the robust peasant that is *M. horridula* and the slightly effeminate *M. latifolia* and is worth persisting with. It is possible with the multitude of seed arriving that more variation in form and tolerance will be forthcoming, since some recent reports suggest that it is found growing in quite dry areas too. A strikingly beautiful white form with purple anthers and stigma turned up from wild seed and such plants are worth much effort to establish.

M. ARGEMONANTHA

A rare and little known monocarpic deciduous species with white or yellow flowers.

This species was imperfectly known when Taylor wrote his book but subsequently he himself examined it in the field and even collected seed. It is a 30 cm (12 in) tall plant of cliff ledges and has distinctive foliage (Figure 14) with extreme lobing such that it appears to have individual leaflets off the leaf stem. It is one of the monocarpic deciduous species but has never succeeded in cultivation, although seed was obtained by Ludlow and Sherriff on their expedition with Taylor. There are two forms of this plant, the white-flowered *M. argemonantha* var. *genuina* and the yellow flowered *M. a.* var. *lutea.*

M. BARBISETA

A perennial blue-flowered species from Qinghai in China found at 4,400 m (14,500 ft).

It is one of four new species described by Chinese botanists since 1952, the others being *M. pinnatifolia*, *M. wumungensis* and *M. zangnanensis*. An account of these new species has been produced by Stephen Haw (1980). It is a plant 30–37.5 cm (12–15 in) high with basal leaves and a single flower stalk. It is apparently characterised by a fleshy root. The 8 or 9 cm (3–3½ in) flowers are quite large with cream-coloured anthers. The plant is presumably related to *M. quintuplinervia* and sounds rather nice, especially as it is perennial.

M. BELLA (Plate 2)

A dwarf, difficult and desirable species with pink or blue flowers, AM 1938.

It should be perennial and is related to species like *M. grandis*. This is the challenging species in the genus among those that have been attempted and it has been successfully raised from seed to flowering in the past, but not, it must be immediately confessed, by the author. It is a perennial species that dies back to a tiny, living resting bud in the centre of the dead rosettes of the previous year's growth. The leaves are all basal and may be simple in some forms on a long petiole or irregularly pinnately lobed. The flowers are borne in succession on long basal scapes up to 10 cm (4 in) in length. They are reportedly very large for the size of the plant and up to 7 cm (2¾ in) across. The flower colour is a powder-blue, a soft pink or a purple-blue. One briefly in cultivation recently, after collection as a plant, was of a disappointing purple-blue and rather knocked one's idea of the species as a fabled holy grail. It is a crevice plant of shaded rocky outcrops and never abundant. It has an exceptionally long tap root system which clearly takes some years to establish and the plants, even when mature and flowering, are pressed

1. *M. aculeata*

2. *M. bella* in the wild in central Nepal (R. McBeath)

3. *M. delavayi* in the wild in Yunnan, China (R. McBeath)

4. *M. gracilipes*

5. *M. grandis*, 'Betty Sherriff's Dream Poppy', at Keillour (H. Taylor)

against the surface only a few centimetres above the ground. The seed is distinctive being elliptical with a sharp point at one end and a very fine reticulation on the surface and of a pale brown compared to other species.

Failing to grow this on to flowering has at least taught me a number of lessons. One plant did survive into its fourth summer and the recipe below if carried out without any slip-ups is probably good enough for success. The immediate need is seed and until 1988 it was available commercially and one must hope that it will again be collected in the Himalayas by one of the Indian wild seed collecting companies. The seed must be thinly sown, in spring, to avoid transplanting of crowded seedlings. It is not transplanting that is fatal, indeed they are fairly tough to handle even with only a pair of minute true leaves, but the fact that it checks them for up to a month and rapid growth is essential in the first year to produce plants that are large enough to survive the winter. The compost mixture is vital with this species and must include at least 50 per cent sphagnum moss in the top 2.5 cm (1 in). The sphagnum should be dried in an oven at about 70°C (158°F) until it is brittle-dry and can be rubbed through a fine meshed sieve. This is mixed with 10 per cent Perlite (expanded rock material) or possibly a really sharp sand (Cornish) in the same proportion and 40 per cent of good quality sieved soil-less peat-based compost. This mix should occupy the top 2–3 cm (1 in or so) and below this, in a pan at least 8 cm (3 in) deep, the proportion of dried sphagnum should be reduced and replaced by leafmould. The dried sphagnum is absolutely vital for reasons discussed earlier (see p. 35) but mainly to produce a humid micro-climate. The seed should be sown thinly on to a lightly pressed surface of a pot with the compost right up to the very rim, and the finest dusting of covering given. Gentle watering is probably normally sufficient to settle the seed. It is much the best thing to sow the seeds individually and ensure the proper spacing and it will be time well spent. What one is looking for is 100 per cent humidity at the immediate surface but buoyant circulating conditions above that.

Once sown the pots should be treated as for all other species. Germination should be rapid and within two weeks. A mist nozzle in the heated frame is excellent since a watering can will deliver far too robust a stream of water. A small hand sprayer is the answer for a few pots.

The seedlings will grow on incredibly slowly even if they are not transplanted and it is at least two months before the roots have extended to 2.5 cm (1 in). The leaves by this time are still only a few millimetres in diameter on a stalk about the same length. A very little feeding with dilute liquid feed is in order on a weekly basis. As long as they stay humid they grow without difficulty at this stage, albeit desperately slowly. Slugs are lethal to them; unless the ring of slug pellets is pretty comprehensive they penetrate your defences. The plants are so tiny that even a small slug regards a seedling as a little snack before starting on an evening of horticultural mayhem. There is no sadder sight than a little ring of leaves cut off at the centre and a slimy trail. *M. bella* are surprisingly tough and they will recover even from this vandalistic treatment, but it sets them back by a month or two and they have almost no reserves in their roots at this time. To help stave off any damage, the frame can be made as slug-proof as possible and plenty of bait laid outside the vital pots to waylay marauders before they arrive. Pots at this time can be enclosed in a large transparent container to maintain humidity and one can leave them for weeks on end like this. This species actually thrives on this treatment, which would damp off most others, and it is no doubt the technique to use in a dry summer spell.

They should, by the end of the summer, have about eight pairs of leaves with some of the earlier ones dying away and the whole plant will be less than 1 cm (½ in) across. Keep the slug bait renewed or the unguarded moment will undo the year's work and there will be no recovery at this time of year. It is now time to change the growing conditions for the winter. The first problem is that liverworts may be growing over the pan. Don't let them start if you can help it because half the compost comes away if they are established when you try to remove them. Mosses can also be a nuisance although a sphagnum compost is usually less troublesome than a peat-based compost. Both moss and liverwort are probably best dealt with by careful hand removal. Scattering a little more dried sphagnum over the rest of the surface regularly keeps the offending growth at bay. Do not use moss or liverwort killers; they kill *M. bella* too.

The compost should remain damp but a little collar of fine grit round the plants is valuable. As cold weather approaches the leaves all die back to leave a bud that is literally invisible unless a microscope is used. This minute gem of green will stay dormant until spring wakes it up. It would appear that in this state it is moderately frost hardy. In the wild however it is probably submerged by snow, and the correct proceedure is to have the pots or pans of seedlings plunged so that the compost will not freeze and have polystyrene chips in bags (I use old onion sacks) that can be laid over the top in cold weather. Plants came through a number of spells below –15°C (5°F) using this technique. There is however one awful problem, at least in the unreliable British winter, and that is a long mild spell. The winter of 1987–8 in Britain was mild over the New Year and the plants of *M. bella* (2 + years old) were all back in growth by mid-January and looking very happy. I had been aware for some time that some species were only frost hardy when dormant and made efforts to protect them against any frost. However, one night in mid-March it dropped unexpectedly after midnight to –10°C (14°F) and, although the plants were covered, this was not enough and the whole collection perished.

The only positive side of this tragedy was a thorough post-mortem of the root growth system of all the seedlings. This showed a 5 cm (2 in) fibrous root system growing out into all directions in the compost but with a main central trunk and the very top of the root was swollen as a storage organ. The answer is clearly to grow them frost free as soon as there is any sign of regrowth and even when dormant not to risk more than a degree or so of air frost without thorough protection. The compost must be at least moist at all times and the great advantage of sphagnum compost is that this is easily achieved without waterlogging. It would be safest to water from below in winter.

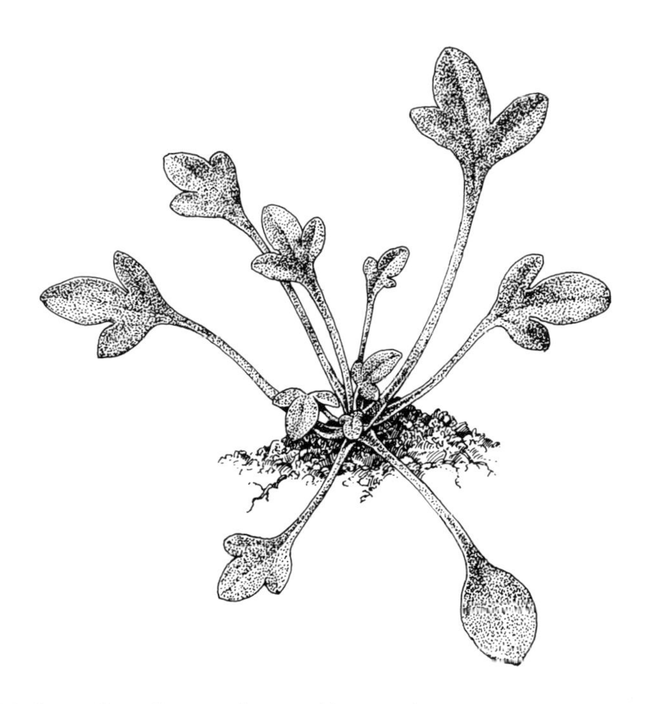

11. *M. bella* seedling (25 mm (one inch) across)

Once in growth at the beginning of the second year they can be potted into a leafmould, grit and peat compost with added inorganic fertiliser or better still some resin-based slow release fertiliser. The top 2.5 cm (1 in) must still contain up to 50 per cent of dried fertilised sieved sphagnum. Slug protection must be continuously effective and all chemicals avoided. By the end of the second year some leaves with their stalks will be nearly 1.5 cm (½ in) long and the whole plant 2.5 cm (1 in) across! Leaves may be simple or lobed and both types are found in the wild. The seedlings I have grown have eventually all produced lobed leaves but sometimes not until their third year. This same growing procedure will need to be maintained for at least two more years. I should guess that some might flower at four years if perfectly grown but with checks they will take longer.

In the wild this species is a crevice plant, often with water dripping on to it, and found at high altitude. People writing with field experience describe a very long tap root penetrating deep into the rock face fissures. The strategy of plants growing in such a hostile environment is clearly to spend a long time with the minimum of leaf foliage, gradually extending the roots deep into the rocky substrate ensuring an adequate supply of food and moisture before attempting to flower. The long tap root will gradually swell to become a storage organ but the accumulation of such reserves is a long process. If seedlings can be matured, a north-facing rock crevice should be contrived in a misted scree covered in winter and a good parcel of dried bracken should be on hand to protect against late spring frosts. This may seem a lot of attention for a plant hardly in cultivation but it has two virtues. In the first place it will always be a small plant. The second virtue is that having had a fairly extensive love affair with this plant I truly believe it wants to please. I have had so many reputably difficult plants die before I could even introduce myself to them that I recognise a plant that wants to co-operate and really has a rather forgiving nature. If a good source of seed was guaranteed for a few more years we could triumph with this. The slight cloud on the horizon is that the Edinburgh plant that flowered would have needed 'special offer' status at a garden centre to ensure a sale. They are described as powder-blue and pink in the wild with extra large flowers so perhaps there are good and bad forms. I fear stage two in the quest to grow *M. bella* may be to grow a good form, but then life is never simple!

M. BETONICIFOLIA

The best-known Himalayan poppy and certainly the easiest of the perennial species, AM 1933.

12. Flowers of typical *M. betonicifolia* (left), *M.* × *sheldonii* (centre) and *M. grandis* (right)

This can be a fairly variable species both in form and colour. It can be soundly perennial and is totally deciduous with only the dead and dried remains of the previous year's foliage marking its winter position. It emerges fairly early in spring with sharp pointed furry little ears of leaves with characteristic dark brown hairs with a purple tinge in some forms. There is a white form that has even darker brown hairs on a pale green background at the same stage in spring and, once one has seen the white and the blue together, they can be told apart with utter reliability. Indeed the colour of the emerging leaf tips in spring is very significant in identifying forms of *M. betonicifolia* and *M. grandis* (see also Figure 7) as well as the hybrid between the two *M.* × *sheldonii*. The hair colour no longer dominates as the plants emerge and the flowering spikes elongate and only the pattern of notching on the leaves and the close set of the stem leaves are characteristic. The flowers usually have short pedicels only 10–15 cm (4–6 in) long and there are normally four to six petals but multiplex forms occasionally occur. The flowers can be up to 8 cm (3 in) or more across in good forms and of the most perfect sky-blue. Some are a muddy colour with various washes of purple and some even come close to pink. There are deep pure purple forms as well as white. Some plants are very much superior to others and if isolated appear to breed true.

There is also the form *pratensis* that is distinguished by very deep purple leaves in spring (indeed it is identical to some form of *M. grandis* at this stage) which is very robust in growth and often approaches 2 m (6 ft) in height in very rich ground. The plant is in all respects identical to normal *M. betonicifolia* and definitely not *M. grandis*. The origin of this plant is a little obscure but it is sometimes called 'Ascrievie form' after the home of Major Sherriff and his wife. Kingdon-Ward introduced a sub-species (KW 6862) which he called 'pratensis' from his 1938 expedition, but it is not clear if the current plants are derived from this or whether wishful thinking at some stage gave this plant its name. It does not set too much viable seed which is slightly suspicious. The flowers of this form are good average but not exceptional.

There is available commercially a form of *M. betonicifolia* known as 'Branklyn'; this is unfortunate since 'Branklyn' is actually a form of *M. grandis* and there is enough confusion about which is the true plant as it is.

Seed is also sold as *M. baileyi* by the trade but this name has long since been superseded by *M. betonicifolia*. *M. baileyi* was described as being distinct from *M. betonicifolia* because of the relative length of the style and the hairiness of the ovary. Taylor concluded, and no one has since seriously argued with him, that they were variations on a single species.

The major problem to overcome with this species is to produce a fully polycarpic collection of plants from seed. If plants are to be obtained already growing it is better to try and obtain a division of an established clump than buy a seedling. There is much advice written about growing from seed but what follows is probably the best solution. Seed should be sown very early in spring as described under general cultivation and then pricked on and later potted into 10 cm (4 in) pots. It is the most likely of the easier meconopsis to damp off and it is essential to avoid sowing too thickly. A dozen seedlings in a

10 cm (4 in) pan will grow better and faster than a hundred of which 80 per cent will be thrown away anyway. The whole essence of success is to grow as fast as possible without making them unduly soft. They should be large enough to plant out by mid-August and, if in a good rich bed, will be 15 cm (6 in) plants before they go dormant. These plants will grow rapidly in the following spring and may flower but they should be strong enough to do this and be polycarpic.

A recommendation described in great detail to me by Alec Duguid, who ran Edrom Nurseries which lie east of Edinburgh on the south side of the river Forth, was to sow seed about six weeks after harvesting and grow the plants on that autumn so that by the end of the following summer they were really robust but had not of course flowered. This is the perfect way to grow them from seed but it does require skill. Starting them as early as possible in spring is probably the best general method but both rely on producing really robust plants the second spring. Plants overwintered in pots from a late spring sowing do not develop a robust structure before they flower that year and are doomed to die. Really poor plants in pots may avoid this fate by failing to flower in the year planted out and then making up for a deprived childhood over the next summer, but this is a shameful treatment. Good rich feeding, no checks and unlimited root run are what are aimed for. It must be remembered when all is said and done, that in any batch a proportion are probably monocarpic and maybe some strains are more prone to it than others. One should expect to establish a minimum of one-third of a planting, even if conditions are not always perfect, and a higher percentage if established in a good, warm, humid spring and early summer.

It has been said that in the wild they grow in sandy soils as well as in open woodland or damp meadows. In cultivation they certainly will grow well on dry sandy soils and more easily than their nobler cousin *M. grandis*. They tend, however, in drier soils to be a less good blue and in a really dry summer, such as we have in Britain every five or six years, they would be badly weakened and may fade away if not die at once. They are particularly vulnerable to mildew (see Appendix V on diseases) in dry weather and, like all plants that become infected, once it is established, it is there for the season. In dry areas there is no doubt that routine spraying every two weeks in spring, with a systemic fungicide alternated with a non-systemic one, will probably prevent the infection. Care is needed with all chemicals on meconopsis as they can cause scorch. Prevention, as always, being better than the impossible cure.

Once established, they are soundly perennial as long as they are kept in rich growing conditions. Without doubt, the finest specimens of this species will be grown in a really rich acid soil with high humidity and light shade; but they will grow reasonably in most seasons in the conditions of an ordinary herbaceous border provided that they are not scorched by a torrid sun and their massive fibrous root ball does not suffer undue competition for scarce summer moisture from less well-bred neighbours. In a dry summer, unless there is high humidity, *M. betonicifolia* will lose more moisture through its leaves than its roots can obtain and the plants will wilt. They are vulnerable to mildew when in this wilted state and any amount of ground water will not restore turgor. The only way this can be overcome is to install a mist unit overhead and to saturate the whole plant habitat night and morning. This is an excellent solution and would probably allow a number of species to be grown in really hot dry summer climates. However, a hot dry summer in Britain usually means water rationing and then it's up to your conscience!

I do feel however that a soil that is alkaline will affect the colour blue and this can lead to disappointment; I garden on a dry limy sand soil and a huge amount of peat produced the electrifying blue this species is capable of. At its best the colour is almost supernatural in its intensity but gradually the alkalinity has drawn up and washes of purple bring streaks of commonplace to the flowers. The answer is a hard pan of peat stamped into a highly compressed layer—maybe 15 cm (6 in) of sedge peat—and then 45 cm (18 in) of a peat-based rich compost on top.

Vegetative propagation of this species is best carried out in early spring when there is just enough growth to see what you are doing. There are two methods: the first is simply to split a good clump in two across a natural cleavage plane. A really big piece bent in the middle normally splits into two without harm. Sometimes during this operation smaller pieces break off and these can be cleaned of dead foliage and inserted in a good gritty compost as cuttings, perhaps dipping them first in a fungicide. The alternative method is to scrape away in early spring a 15 cm (6 in) circle around the plant and fill it with a rich leafmould compost. The plant will send offshoots out into this which will then root freely. The following spring these can easily be removed and grown on. This is probably the best way to produce large numbers of plants of a good strain. It is not recommended to split large clumps up into lots of pieces.

M. CAMBRICA

The perennial wildling with yellow flowers from western Europe.

This is the only non-Asian species and the plant that originally defined the genus. It is an accommodating plant with a deciduous perennial habit growing both in shady woods and dry crevices in walls. It is a dwarf plant less than 30 cm (12 in) high with dissected smooth leaves and delicate yellow flowers borne on basal scapes. It is distinguished from other yellow poppies by having a style between the seed capsule and the stigma. There are orange forms and a deep orange form recently named 'Frances Perry' (it is not scarlet!). Both colours are present in multipetalled forms, which I find ugly. In the garden they need regular dead-heading or the widespread seedlings will become a nuisance. They can be awkward to raise from seed and the best recipe is to sow them directly in a finely worked soil where they are to flower, preferably in autumn.

M. CHELONIDIFOLIA

A perennial yellow-flowered species from China.

The flowers closely resemble those of *M. cambrica* but the plant is straggly and over 1 m (3 ft) high. It is adapted to be supported by the woody vegetation of thickets. The leaves are both basal and on branched flowering stems and are usually lobed into three. It has the habit of producing vegetative buds in the upper leaf axils which can be grown on. It also produces offsets on rootstalks which root and provide further material for propagation. This is just as well since it rarely seems to set seed. It may be that there only a few sterile clones in cultivation or, like many other species, it is self-sterile. Ratter (1968) commented in his study of *Meconopsis* chromosomes that this species had little viable pollen in plants he examined. The related *M. villosa* does however set viable seed on self-pollinated plants. It is a plant for damp shaded sites in the wild garden and will accept very poor soil conditions. A worthy but dull relative in a family of superlatives!

M. DELAVAYI (Plate 3)

A rare, difficult and desirable perennial species probably hovering on the edge of extinction in cultivation, FCC 1913.

This is a species with simple rather glaucous deep green leaves, and large slightly pendulous violet flowers born on a basal scape. The flowering stems are about 25 cm (10 in) in a good specimen and twice that at fruiting time. The leaves are all basal and borne on a petiole 7.5 cm (3 in) long and are only sparsely covered in spines. The texture and simple shape of the leaves is not typical of the furry or spiny dissected leaves of meconopsis which is why this species is placed in its own separate series. The flower is however quite typical of the genus, as is the seed pod which is distinctively long at up to 7 cm (3 in). The rootstock is branching and swollen as a storage organ with a main tap root 25 cm (10 in) or more. The plant is deciduous and should be reliably perennial.

The plant has always been uncommon in cultivation, but R.D. Trotter of Brin House in Inverness mastered its cultivation and grew this plant extremely well. In the wild it is found on limestone, as are some other species, but it is not essential. Trotter used a sunny and a half shaded aspect equally successfully. His significant discovery is that this species can be propagated from cuttings and root cuttings. Trotter accidentally broke off a tap root 20 cm (8 in) down. Two lengths from the broken top were successfully rooted as cuttings. The bottom part of the tap root also regrew. The plant thus resembles a number of other herbaceous plants such as the red *Papaver orientalis*, where sections of root, inserted top upright in a sandy compost, will throw shoots and develop a full root system. It would be worth trying this in late spring, using bottom heat if new material can be collected in China. It has been suggested that it is difficult to prick on from seedlings; while all species are susceptible to rough handling at this stage, great care should be taken until we know more of this aspect of growing them on from seed. This species grows in regions of China that are now being reached by Western gardeners and indeed some seed has been obtained. It is therefore likely that this species could be re-introduced in the near future and, with a new generation of expert plantsmen, there is little doubt that this species should not prove unduly difficult if Trotter's success is built on.

M. DHWOJII

An easy and attractive monocarpic species with an evergreen rosette.

This is a rather delicate species which flowers in two to three years from seed, gradually building to quite a substantial rosette. A flowering spike emerges and grows to about 1 m (3 ft) and bears graceful panicles of cup-shaped pale yellow flowers about 4 cm (1½ in) across. There are normally

13. *M. dhwojii* (leaf)

two to three flowers on each panicle. The plant is readily distinguished from all other rosette-forming species in cultivation by the substantial amount of purple pigment associated with the leaves and spines. The leaves are highly dissected and lobed giving the rosette its very delicate appearance. This species requires standard cultivation treatment recommended for all the monocarpic rosette-forming species.

It requires a rich open soil and some shade is desirable except in very northerly or very humid areas. I am certain that an acid soil is not essential as long as there is much organic matter in it. It is not especially susceptible to winter crown rot and is quite satisfactorily overwintered uncovered if there is good drainage. The necessity of covering the evergreen rosette species is something that may have to be worked out for a particular garden situation, but *M. dhowjii* is not a difficult species. It is related to *M. gracilipes* which is only distinguished by the rather superficial characteristic of lacking the purple pigment of the leaves and spines of *M. dhwojii*. Both these species cross-pollinate with the *M. napaulensis* types and the progeny are sterile. This cross is known as *M.* × *ramsdeniorum* and has no great merit. In garden situations *M. dhwojii* should be kept as separate as possible from related species if viable seed is to be obtained.

M. DISCIGERA

A difficult and unusual monocarpic species that has been introduced and lost a number of times.

This species is one of two (*M. torquata* being the other, although there is now, possibly, a third species *M. pinnatifolia* described by the Chinese) that Sir George Taylor placed in a separate sub-genus. They are distinguished from all other species by the presence of a flat disc surmounting the ovary although they are unmistakably meconopsis. This is a monocarpic species that dies back in winter to a moderately tight but large resting bud. It has been suggested that this resembles a bird's nest.

I should declare at this stage that I have tried to grow it twice from seed but, although I have kept it alive for four years, I have not flowered it. It forms a typical basal rosette from which long narrow strap-like leaves arise. These can be up to 20 cm (8 in) long including the linear petiole and in all the plants I have seen they have a very characteristic tri-lobed tip which makes the leaf quite distinct from any other species. Taylor implies the leaves can be simple without the notching and forms with dissected leaves have been described in the wild. The whole rosette gradually increases in size with a dense mass of leaves. Many of these die each autumn and the new ones in the centre are largely unfolded. The leaves expand and the rosette gradually becomes larger, although in four years it still may be only 18–20 cm (7–8 in) across with a mass of dead leaves around the outside from previous years. It finally develops a flowering spike up to 45 cm (18 in) in height. This has up to 20 flowers on it with short flower stems which open from the top. The flowers are fairly compact in much the same way as good specimens of *M. horridula*. Taylor describes the flowers as blue, red or purple. He mentions one that flowered yellow in Edinburgh in 1930. It has subsequently become clear that yellow *M. discigera* are widespread and such coloured plants have been illustrated in the wild. One plant that flowered from the Alpine Garden Society's Sikkim expedition also flowered yellow. I find it interesting that this was not a bright yellow but the pale cream yellow of blue/yellow crosses in the genus (i.e. like *M.×beamishii*) and more field information is desirable for this widespread species.

14. Leaves of *M. discigera* (left), *M. violaceae* (centre) and *M. argemonantha* (right) (not to same scale)

There is no doubt that it is difficult. A great deal of excellent seed was brought back recently from Sikkim, but I think few have been brought to flowering. It requires the three basic ingredients for difficult meconopsis—a dry winter, a moist summer and feeding; ways of creating these have been described in the general section on cultivation. Great judgement as to when to send it dormant in autumn is needed and all my losses have been at this period from a black fungal rot that starts right at the base of the rosette, and the plant has gone before you detect any symptoms. It is especially vulnerable, before it has died back, to damp mild spells in autumn. My main successful germination was in 1984 which was a tryingly hot summer in the

UK for meconopsis and it was succeeded by a hopelessly wet year right into the autumn. These climatic conditions were both disastrous and killed off most of the plants before I had begun to understand the requirements. The one plant that I know of that was brought to successful flowering was grown close to a hedge. Anyone who has hedges in the garden is well aware that by late summer the huge moisture demands of the roots combined with a thick canopy of summer growth produce an extraordinarily dry micro-climate even in really wet weather. A modification of hedge conditions may well be the answer—a rich bed overhung by a spreading juniper and the difficult *Primula forrestii* flowers annually for me, totally neglected, in this situation.

M. FORRESTII

A plant related to *M. horridula* with pale blue or purple-blue flowers best distinguished by long narrow fruiting capsules.

This has occasionally been in cultivation but has not persisted. It will require the same summer moist and winter dry growing conditions as is suggested for species like *M. speciosa*. Seed of this plant has recently been brought back from China and plants are growing well although the identification will have to be confirmed when they flower. The leaves have few golden hairs and are a glaucous green and the plants have produced a typical tap root after the first year of growth. They become winter-dormant much later than related species with leaves present into November. I am exercising great care with slug control and they are being kept winter dry. They are certainly large enough to flower in spring 1989 and this seems likely to be for the first time in cultivation.

M. GRACILIPES (Plate 4)

An especially beautiful monocarpic evergreen rosette-forming species with yellow flowers.

This species now seems to be very rare in cultivation. It is a very close relative of *M. dhwojii* and only distinguished by the lack of purple pigment at the spine bases. It takes from 2 to 4 years to flower from seed and gradually develops a large rosette up to 75 cm (2½ ft) across or more in a really well-grown specimen. The deep rich green leaves are very finely dissected giving a most delicate fern-like appearance. The flowering stem can be 125 cm (4 ft) high but in general forms a graceful pyramid of flowers of excellent proportions. The flowers, which can be up to 5 on a pedicel, are not large being 6 cm (2½ in) at most and of a neat cup shape. Although only a small number of flowers are out at any one time it manages to remain a tidy-looking plant long after one is desperately waiting to haul out the dying remains of related species such as *M. napaulensis*. It does not appear to be any more difficult to grow than *M. dhwojii* and *M. napaulensis*, given good rich feeding and a moist site. It may be more susceptible to winter wet and, as it is now rare, it should perhaps be given the benefit of the doubt and given overhead cover.

The main planting of this species that I enjoyed flowered over two years but set no seed from dozens of plants; they did not appear hybrid and their sterility is a mystery. I have since obtained seed from two other sources and

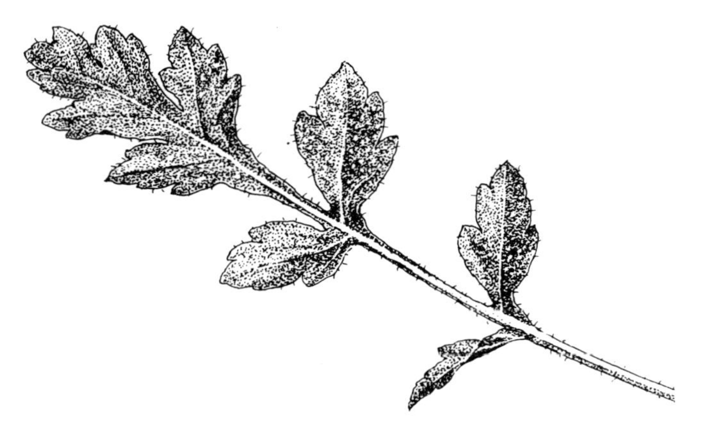

15. *M. gracilipes* leaf

both were undoubtedly a hybrid with *M. napaulensis* showing all the coarseness of the cross and were sterile. Seed of this species has not been offered lately and unless it has been re-introduced may no longer be in cultivation. If seed of this species is obtained care must be exercised to keep it from hybridising with any of the related species. It is very desirable, especially set in a woodland glade, where it would provide all the grace of an evergreen fern with the added advantage of finally expiring in the most elegant fountain of refined flowers.

M. GRANDIS

One of the classic garden plants worth any effort, a perennial with breathtakingly blue flowers.

M. grandis is unmistakable but is better illustrated (see Figure 9) than described in words. Nearly all seed offered as *M. grandis*, and many plants too, are in fact *M. betonicifolia*. Much *M. betonicifolia* seed comes into the various seed exchanges, often under various disguises, but there are very few cultivated plants of *M. grandis* that yield seed. There are many named forms and clones of *M. grandis* and there is much confusion about these names; some are undoubtedly hybrids with *M. betonicifolia*. The true species is very easy from seed and normally grows away strongly. Adult plants are usually clump-forming with flowering and non-flowering rosettes in each season. They become fully dormant in winter though in an untidy garden the dead leaves and stems remain overwinter. The leaf buds emerging in spring are normally purple but the Sikkim form, for example, is a pale green as it emerges, covered with straw hairs, which hardly colour it. The leaves rapidly form a clump up to 60 cm (2 ft) across and 45 cm (18 in) high from which a

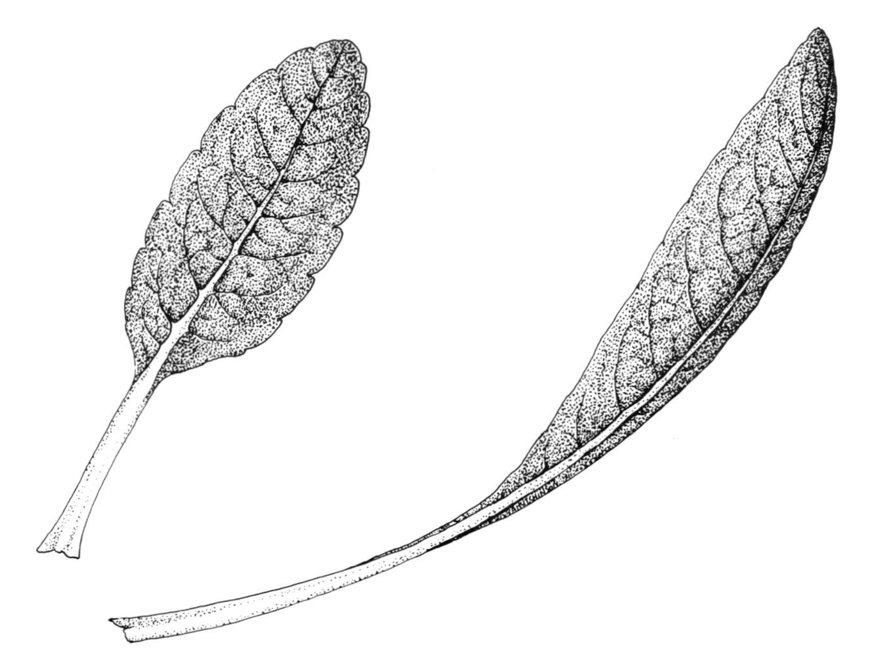

16. Comparison of narrow leaf of Sikkim form (right) with broader leaf of Nepal form (left)

number of flowering stems arise. These flowering stems usually have at least one whorl of stem leaves, though single flowers on basal scapes can occur and are typical of some forms of the plants. Basal flowering scapes can be associated with weak seedlings flowering in their second year and often provide the decoration for the funeral of the plant! The flowering pedicels are usually about 15 cm (6 in) long but can be three times that length.

The flowers are blue, white or purple with many shades in between. The most common blue in *M. grandis* is a dark blue tinged with purple. It is rarely the pale blue of *M. betonicifolia* or the green-blue that electrifies the viewer in perfect *M.*× *sheldonii*. There is no doubt that a number of different growing conditions affect the colour and I now believe that pH is the most significant factor. I have watched all my named forms and hybrids gradually become more purple with only the hybrid *M.* 'Houndwood' having the decency to put up a fight. The common factor is a naturally alkaline soil intruding into the prepared peat beds. The purple colour is reversed when replanted in acid soils. Chemical treatments for turning the flowers to a good blue do not seem effective. The making of a blue poppy habitat is discussed on p. 20. There is little doubt that this species is less tolerant of hot dry sun than *M. betonicifolia* and is thus potentially more difficult in climates with a typical continental summer. It is possible that there are some forms that are more drought tolerant than others, given the variable habitats and types of this species in the wild, but very little has been done to select such forms.

The biggest problem with this species is the same as that of *M. betonicifolia* when grown from seed: many plants will tend to flower in the second year, produce a single stem and then die (see p. 53). In summary, very early sowing is required to produce very strong plants at the end of the first year. Taking out a flowering spike in the second year if done early enough may help but is by no means infallible. It must also be remembered that some strains are more inclined to be monocarpic, and there will be a few such plants in any batch of seedlings. It goes without saying that seed should be saved only from very good forms but this is usually easier said than done because many good forms appear sterile.

There are a number of forms of *M. grandis* in cultivation. Some of these are based on collectors' numbers, though it is far from certain that this is their real origin and, of course, in some cases subsequent generations have been raised in cultivation under these collectors' numbers which is an inappropriate use. It is at least possible that some of these forms are hybrids anyway and are really forms of *M.* × *sheldonii*.

'Alba'.
There are lovely white forms but they are rare and difficult to obtain.

'Betty Sherriff's Dream Poppy'. (Plate 5)
Collected by George Sherriff's wife after dreaming where to find this plant; the story is recounted in Fletcher's book, *In Quest of Flowers*. Plants of this still exist and set seed.

B.M.W. 109.
A form collected in Nepal by the expedition led by Binns. Tends to be purple-blue but seeds reliably.

'Branklyn'.
A selection made from GS 600 seedlings by the Rentons of Branklyn Gardens. It is now not quite certain what this plant was, as Branklyn Gardens have recently re-collected a number of different plants under this name. It seems most likely to be a very tall plant with very large good blue flowers.

GS 600.
Perhaps the most notable collection of Sherriff. A marvellous clear blue but it rarely sets seed and, although it is offered vegetatively propagated, one

cannot be sure if these are the original clone but they do fit Sherriff's description and were grown until very recently in their garden at Ascrievie.

'Ivory'.
Plants under this name are in the national collection at Durham.

'Keillour Crimson'.
A strain from Nepal that has deep purple flowers (crimson is wishful thinking). It was named and exhibited by the Knox-Finlays of Keillour.

LSH 21069.
Plants still exist with this number from a Ludlow, Sherriff and Hicks collection but one cannot be certain they are the originals.

'Miss Dickson'.
A beautiful white form but probably it is a hybrid of the *M.* × *sheldonii* type. I have seen a plant but it is very rare.

'Miss Jebbs'.
A very desirable dwarf deep blue form with cup-shaped flowers. It flowers at less than 1 m (3 ft) and remains compact. Safely in cultivation but not common.

'Nepal Form'.
The most widely available form with large flowers but usually of a purple-blue. This set seed at one time but now rarely seems to.

'Prains Form'.
A deep purple from that was mis-named and subsequently renamed 'Keillour Crimson'. 'Slieve Donard' was also briefly known by this name.

'Puritan'.
A form grown in the national collection.

'Sikkim Form'.
These are early-flowering with very narrow leaves of a plain green in spring. This form at present sets seeds. The flowers are generally smaller than average and of purple-blue.

M. HENRICI

A Chinese species related to *M. horridula* that has briefly been in cultivation and described by Taylor, based on herbarium specimens, as a desirable garden species.

It commemorates Prince Henri d'Orleans who collected it near Tatsienlu

6. *M. horridula* in the wild (high altitude form), (R. McBeath)

7. *M. horridula* (dark blue form)

8. *M. integrifolia* flower

9. *M. integrifolia* rosette

10. *M. lancifolia* in the wild in Yunnan, China (Peter Cox)

11. *M. latifolia*

12. *M. aculeata × latifolia*

13. *M. napaulensis* (pink form)

14. *M. napaulensis* (blue form)

15. *M. punicea*

in the last century. It was flowered in 1906 from a Wilson seed collection and plants were also cultivated from later Farrer and Kingdon-Ward collections. It did not remain long in cultivation. It is distinguished by having dilated filaments to the stamens. This means that the pollen-bearing anthers are supported on fattened rods and thus some botanical knowledge is required to make this identification. The flowers are normally borne on basal scapes in the original species and sometimes there is only one flower on a plant. The leaves are up to 15 cm (6 in) long and are clearly similar to those of *M. horridula* and variable. The flowers are borne on long basal scapes up to 50 cm (20 in) in length and the stems are covered in rufous bristles. The petals and the stamens are deep purple to violet and the anthers orange or pale yellow.

It was found in alpine pasture and scrub over quite a substantial altitude range so its obvious difficulty in cultivation is not easy to understand. There are no deep purple species in cultivation and the purity of purple in these relatives of *M. horridula* from western China is as desirable as the purity of the pink, blue, yellow and red of the other species we cherish in our gardens. There are forms of its relative *M. horridula* that are difficult and in this case overwintering is the problem. A good layer of dry leaf litter under a small cover is probably a starting point for the winter protection (see also p. 25), and great attention paid in spring that the new growing points are not subject to slug predation.

M. HORRIDULA (Plates 6 and 7)

The one monocarpic deciduous species worth trying anywhere and brilliant in its best forms.

This species is cursed with an unfortunate name; although the prickles are indeed horrid and gathering seed in autumn is not for those with delicate skin, there are some truly lovely forms of this accommodating plant. It is very widespread in the wild both in terms of its spread across the Himalayas but also in altitude and habitat. Taylor lumped together a number of species which indicates that it is very variable but over a continuous spectrum. The plant can have a tall elegant flowering spike with up to 50 dark royal blue flowers without a trace of purple, or it can be a coarse plant with heavy purple-blotched leaves and meaty-mauve flowers. The leaves are nearly always strap-like and smothered in stout spines (though Ludlow and Sherriff described a form with dissected leaves in Tibet). The flowers open from the top of the flowering spike and may be blue in almost any shade from silver to deepest royal blue, with many muddy intermediates as well as pure white, yellow, pure pink and chocolate. The last three colours have not been in cultivation, only described from the wild. The most beautiful light blue forms recently described from high altitudes in Nepal and Sikkim are not easy to grow.

If well grown the plant is normally biennial but winter-dormant to a resting tap root and often even starved plants not potted on will produce a few flowers and die. A small number in a large planting will wait to flower in the third year. It germinates well from seed, though the seedlings are rather gangling with the leaf petioles rapidly expanding and a newly pricked

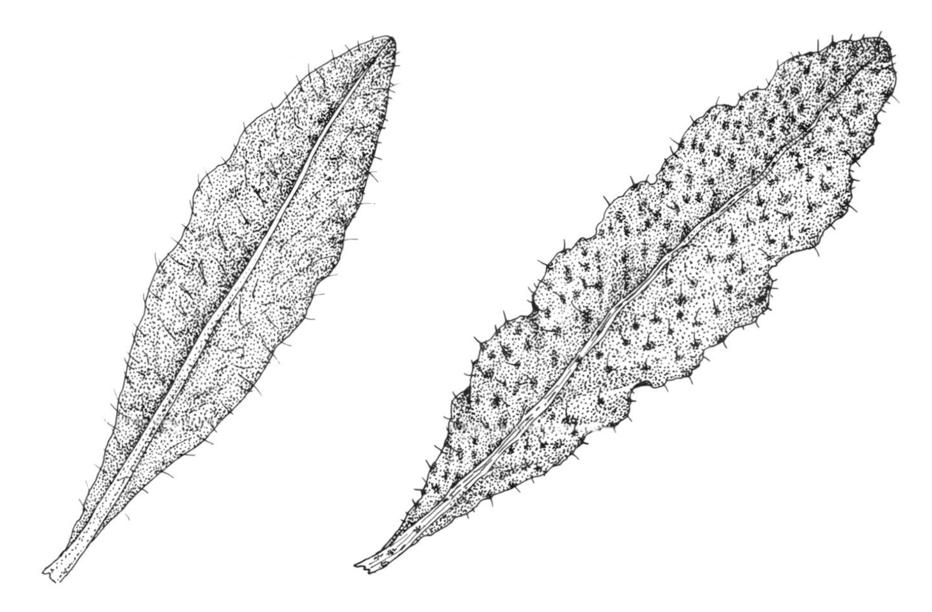

17. Variation in leaves of *M. horridula*; some are coarse with blotches of purple pigment in this very widespread and variable species

on box always looks a bit of a mess. The next stage in their careers at potting on is much the same as they are beginning to form a tap root, and don't come out of the pricking on box in a neat fibrous root mass as do the rosette-forming species. Do not despair, however, they do not seem to object to transplanting as long as it is done when they are around 5–8 cm (2–3 in) in size. It is essential not to check them at the pricking on stage and the long floppy leaves are vulnerable to wind damage. The planted out seedlings will go dormant early in the autumn and well before the related *M. aculeata* and *M. latifolia*.

The flowers open from the top and they can become rather ugly as one waits for the last flowers to finish at the end of June or early July. It is possible to keep the top pinched off and I usually root out all but a few of the very best plants to be saved for seed. Taylor suggested that pinching out the centre early on would produce a better flowering plant in the garden. There is no doubt from the observation of my own such damaged plants that this might be worthwhile though I confess there is something awful about the deliberate mutilation of nature purely for aesthetic reasons. They normally set masses of seed so only a few pods are usually required. There are two good reasons for pulling plants out after flowering, apart from one of general garden tidiness. The main one is that they can be a thorough pest if left to seed; an innocuous weedy little adolescent thing with a few floppy leaves in autumn becomes an overweight bully the next summer, growing over an innocent group of seedling *Nomocharis* or somesuch. The other reason is that it is essential to keep only the very best forms. This species grows in the gravel of the path and even on the pure gravel of the flat roof of part of the house where in desperation containers of spare plants are housed. It responds to severe drought by flowering as a very poor specimen but has a great will to live. I suspect the difficulty in a very dry climate would be keeping the seedlings growing long enough in spring and summer to attain a viable size.

Euan Cox of Glendoick who wrote the cultural sections in Taylor's book (Taylor 1934) commented many years ago that conventional wisdom grows this species in a rocky scree where they undoubtedly will be quite satisfactory. He was in no doubt that they could be magnificent when grown in rich humid conditions. I tend to grow them in both situations, but the best plants are always in the good rich soil especially in full sun. They only expand for a brief period to occupy very much of such a site and are delightfully easy to pull out (rather easier than a carrot except for the prickles) as soon as they have flowered. There are pinkish, white and blue forms, without names, in cultivation. The colour in this species does not seem to be affected by soil as with the *M. grandis*/*M. betonicifolia* group and in general they breed true once a strain has been selected. The pink forms have all been a rather purply-pink but it might be possible with care to fix a pure form and these have been reported in the wild.

There is also a yellow form which was found in the wild by Taylor on the expedition he joined to Bhutan with Ludlow and Sherriff in 1938. This blue/yellow phenomenon I find interesting as has been discussed earlier. In all cases the cross between blue and yellow is cream in meconopsis. This is seen in all the hybrids with *M. integrifolia* (with four different blue species). The same cream occurs in the cross between blue and yellow *M. napaulensis*. One might expect then to find cream *M. horridula* in the wild if the yellow and blue form overlap but this has never been reported. The genetics of these plants is clearly interesting but has not been studied. The yellow form is not in cultivation.

One of the most beautiful forms of *M. horridula* is the pure turquoise dwarf form from high altitude Sikkim and Nepal. This plant was brought back as seed by the Alpine Garden Society Expedition to Sikkim in 1983 and also by Ron McBeath of the Edinburgh Botanic Garden from the Makalu region. It has proved very difficult to cultivate. Initially seedlings grew quite well but were of small size at the end of the first growing season. Many of the seedlings did not survive the first winter although the reasons for this are not clear since it seems unlikely that frost hardiness could be a factor. This plant is clearly highly desirable both in colour and dwarfness. It is difficult to offer any sound guidance on how to approach this variety, except to suggest that the really critical stage is the changeover from summer growth to winter dormancy and that spring should not come too early.

This species requires and deserves some attention to produce strains that are really well proportioned with flowers of a good colour or range of colours. Sir George Taylor commented many years ago that they were variable and often poor and this has devalued them. The white form at present is poor but several generations of careful selection would improve its form. One has only to look at the exquisite white *M. aculeata* in cultivation to see the desirability of attempting this goal. White meconopsis may come second to blue meconopsis but its still an impressive performance.

M. IMPEDITA

A monocarpic relative of *M. horridula* with dark violet flowers, not in cultivation.

This species has been fleetingly in cultivation a number of times and is probably fairly desirable. It has similar leaves to *M. horridula* but not so prickly, and it flowers with single blooms on basal scapes and is up to 30 cm (1 ft) at flowering time, elongating by half as much again in fruit. Taylor showed it to be closely related to *M. pseudovenustra* and *M. venustra*. The differences between these species seem quantitative rather than qualitative and it is difficult to make sensible suggestions as to how to distinguish them. Taylor also notes that *M. lancifolia* in some forms (like some high altitude *M. horridula*) only flower on basal scapes and is also very difficult to distinguish. Recent colour slides of *M.lancifolia* in the wild show how desirable these deep purple species are. There must be the prospect of one of them being introduced in the near future and the suggestion for growing them would have to be based initially on what has been suggested for another purple species, *M. henricii*. It is a plant from drier regions.

M. INTEGRIFOLIA (Plates 8 and 9)

Farrer's 'Lampshade Poppy' and a magnificent yellow monocarpic species becoming winter-dormant and a species worth trying in any garden.

This is another widespread and variable species, quite stunning when good forms are well grown. It has large oval hairy leaves that can be the most delicate pastel blues, pinks and golds when they emerge in spring. Some flowers emerge from basal scapes but most elongate from a central flowering stem. The flowers open upright in some forms and are pendulous in others, have reached 23 cm (9 in) across in cultivation and are reported a fabulous 28 cm (11 in) in the wild. It is usually easy from seed which germinates freely. The seedlings grow more rapidly than any other species and need pricking on in two to three weeks. They need potting on fast and can be planted out in a rich humid bed by mid-July. The degree to which they go dormant depends on the particular strain and this can have very serious consequences for flowering, as once back in growth in spring some are not very hardy.

The strain commonly in cultivation has plain green leaves and the winter resting bud is usually slightly expanded within the dead remains of the previous year's leaves. These come back into growth in early March and

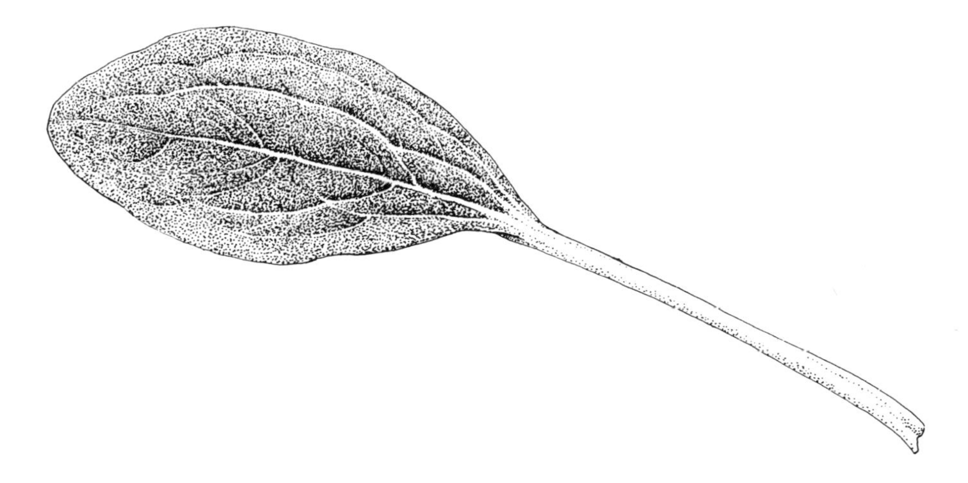

18. *M. integrifolia* leaf

both the flower buds and the young leaves are vulnerable to more than a degree or so of frost. The result is a distorted plant with aborted flower buds that looks as though infected by a virus. The plant dies after flowering and in this case no seed would be set. It is possible to protect them with dried leaves under a simple tent cloche and, in areas where late spring frosts are regular without snow cover, this is probably a sensible precaution at least with a proportion of the plants. The rosettes in spring that expand are quite attractive with a soft covering of pale straw hairs. The leaves are simple and up to 37.5 cm (15 in) in length and often slightly rolled with a prominent keeled midrib underneath. A flower spike then expands with blooms arising from the axils of the upper stem leaves and sometimes from lower down, including an occasional one on a basal scape. A good garden plant has up to ten blooms but often with smaller plants there may be only two or three.

Two recent Chinese collections, one from the Kunming Botanic Gardens collected from an unknown wild site and more recently a collection by Cox and Hutchinson (C.H. and M. 2590) are lovely plants over 1 m (3 ft) high with 15 blooms up to 22 cm (9 in) across. This form has two further great virtues with an exquisite spring rosette and a much later flowering which may avoid frost. These Chinese forms become fully winter-dormant and the dead leaves disappear over winter. The rosettes that emerge from the ground are a soft pinky-yellow or an iron blue-grey and are especially attractive.

If well grown these plants are always biennial but in a casual planting some will not flower until the third year.

It is often said these days that *M. integrifolia* does not flower like it used to. I suspect that when the excellent Chinese forms become better known there will be much nodding of heads and 'I told you so' conversations but I fear that the truth is that familiarity bred contempt: recently Farrer's 'Lampshade Poppy' has been badly grown, rather than the stock deteriorating. It is perfectly possible to grow the strain that has been in cultivation the last 30 years or so to exactly the same size. There is no meconopsis that is greedier than this species nor more rewarding of a good diet. The Chinese strain is good and probably grows to a really striking size more readily, but the main virtue is the spring rosette and especially the greater hardiness in spring. In the past some specimens have been reported as flowering more than one year. This group of meconopsis has a tendency to be perennial and, given the widespread range of this species in the wild, it is perhaps not surprising that these have turned up in the past.

Taylor deals at some length with the variability of this plant and the attempts made by the botanical splitters to create further species. It is very widespread with greatest abundance in north-west Yunnan. Two forms have even been described in the same area, one in scrub and a quite different one in pockets of damp black peat. There have been white forms rarely mentioned and Rock is reported as having collected a pink-flowered form. One can't help wondering however whether the pink-flowered form was not *M. sherriffii* which at that time had not been described. Rock's collecting area was not far from where Sherriff first described his pink poppy and a good description of *M. sherriffii* would in fact be a pink form of *M. integrifolia*.

This species is probably quite possible in dry climates if seedlings are grown on in humid rich frame conditions to a large size and planted out in late autumn. They should come into growth during the brief spring of such climes and flower before the heat of summer dries them out. They might not set seed in such conditions but plenty is usually available. The drier the climate the more shade they would tolerate and indeed would probably be happy with full shade. It is found in the wild at a great altitude range, from 2,700–5,200 m (9,000–17,000 ft), and from rocky screes to alpine meadows, and like the equally widespread *M. horridula* this is sound evidence that this is a species adaptable to many habitats; even in marginal growing areas persistence would probably be rewarded especially if seed from as many provenances as possible was tried.

M. LANCIFOLIA

Another purple-flowered relative of the monocarpic *M. horridula*, not in cultivation but clearly desirable (Plate 10).

The foliage on some specimens is very like that of *M. horridula* but is pinnately lobed. The leaves have a variable amount of bristles and are up to

25 cm (10 in) in length in a strong specimen but are normally all basal. This species has either a flowering scape with relatively few flowers or flowers with single blooms on basal scapes. The flowers have variable numbers of petals, but usually 4–8, are violet or deep purple but Taylor reports light blue very rarely. People used to purple forms of *M. grandis* or even *M. horridula* should not confuse these wan things with the deep rich regal purple of this species. Admittedly such deep-coloured flowers are not easy to site in the garden but if we can do it for the black form of *Iris chrysographes* we could do it for this. This is a species that one may have some hope of obtaining from China because it is widespread in occurrence.

This wide distribution has given rise to many forms and this is another species that Taylor had difficulty in sorting out and once again took the wise course and put a number of different types of plant under a single species. The robust form, previously known as *M. eximia*, Taylor considered as very desirable but he also reduced *M. concinna* and *M. lepida* to forms of *M. lancifolia* and only *M. henricii* survived as a separate species because of its consistent expanded filaments. *M. lancifolia* does not appear to have ever been in cultivation despite substantial efforts by Forrest, which is ominous. Taylor considered that the bulbous root and its generally weak characteristics implied it was a difficult plant. The only optimistic note that one can strike is that it is widespread which shows at least that genetically it is pretty variable and that perhaps some forms will be more adaptable than others. In the wild it is one of the meconopsis particularly associated with limestone, from screes and meadows with less sunny exposures. I suspect the standard well-fed, winter dry, summer wet might require some modification with less emphasis on the summer wet.

M. LATIFOLIA (Plates 11 and 12)

A neglected monocarpic gem with exquisite duck-egg-blue flowers related to *M. horridula*.

This rather scarce species could be regarded as the finest species in cultivation—a suggestion previously put forward by Taylor. Interestingly, Taylor also commented that it had not become as common in cultivation as one would expect. It seeds as freely in my garden as *M. horridula* and is at least as easy to grow in most respects. It is certainly happier in dry conditions than any species except *M. horridula* and receives no special treatment, and yet I have almost never seen it in cultivation elsewhere. I can only assume that it is one of those brilliant garden plants that has simply escaped notice and major efforts should be made to establish it more widely.

The plant is similar to *M. horridula* in growth but is always distinguished by the broader paler green leaves and softer more genteel spines. It differs from *M. aculeata* in that the leaves are never lobed. In the seedling stage, the leaves are rounded and unlike any form of *M. horridula* but there is great similarity to *M. aculeata* until the latter throws leaves that show lobing and this may not be until the 6th or 7th true leaf. This species, like its relatives, becomes fully dormant in winter but the emerging rosette in spring is a delicate silver-pink thing compared to the coarser *M. horridula* and it is later coming through as well. After emergence, however, it is damaged by severe frost. It normally will flower as a biennial. The flowering spike in a well-grown plant can be over 1 m (3 ft) high with the most perfect cup-shaped duck-egg-blue flowers opening from the top and thickly covering the stem. The flowers are long-lasting and although as the flowering period progresses the seed pods form at the top of the spike they are not unreasonably untidy. The plant almost invariably dies after flowering.

A darker blue form that has set viable seed has appeared in the garden in the last two years but it must be at least possible that it is a hybrid. In general, although *M. latifolia* grows mixed up with *M. horridula*, they do not appear to interbreed. There is however a puzzling race of hybrids with variable coloured stigmas that may have one parent *M. latifolia* and the other *M. aculeata*. These hybrids are very attractive because, although the plants are all mauvy-blue, it is that wonderful rippling colouring of shot silk. In general these plants produce very little seed which is further evidence for hybridity. A chance seedling that self-sowed underneath such a hybrid may well have been a back-cross to *M. latifolia* because it was the most perfect pale sky-blue, just one wash of colour deeper than *M. latifolia* itself (Plate 12). The species are so good in themselves there is no point in hybridising them and, as they are monocarpic, they should probably be grown apart since hybridity in this genus often brings first partial and then full sterility. Taylor reports white forms in the wild but they are not in cultivation but as white forms of the two related species are at present rapidly circulating in cultivation it may be only a matter of time. White forms of the blue poppies are usually especially beautiful because they are more compact, though regrettably this does not apply to *M. horridula* at present.

This is one of the western-most species and, as one might expect, is apparently tolerant of drier conditions. It grows happily in moist rich soil and some shade but that is certainly not essential. It needs good feeding and a dry starved hot scree would not be to its liking. One must exercise some caution however since even at the dry end of the Himalayas there is some precipitation at very high altitudes for a lot of the time even when the plains below are burning in shimmering heat. It may well be that it flowers in what appears to be dry conditions but that the strong tap root has its tip in moisture. Seed sowing and subsequent treatment is as usual. It is by reputation tricky at the seedling stage, but I have not found it to be so.

The one real risk with this species occurs in the spring. The genteel bristles of the new growing point from the carrot-like overwintering tap root are the objective of marauding slugs. A few doses of slug pellets are usually sufficient to provide a fatal diversion—both for plants in pots in the frame and plants in the garden, indeed the former are much more vulnerable because of their concentration—but don't delay their application, or you may wait forever for the plants to come through!

In many of the drier climates of the world it should be possible to grow this species in pots in a humid shady small frame and plant them out in the autumn. It is certainly a species to experiment with, even if you cannot grow the classic *M. grandis*, and any success with this exquisite coloured and mannered species is worth the effort.

M. LONGIPETIOLATA

A rare relative of *M. napaulensis* with yellow flowers.

Seed often circulates but is never (so far!) true. This has presumably been in cultivation quite often because it is in several botanic garden catalogues but it is doubtfully in cultivation at present. It is a plant from Nepal so it could return at any time from a wild collection. It resembles a dainty *M. paniculata* (which in turn resembles a dainty *M. napaulensis*) and is distinguished, as its name suggests, by its unusually long leaf stalks. It has highly incised leaves and could I suppose be confused with *M. gracilipes*. It is also distinguished by the lack of a substellate pubescence to the leaves. This is not as helpful as it sounds since even with the benefit of a scanning electron microscope I still cannot reliably find this mysterious pubescence on wild collected material of *M. paniculata*, Figure 21 (see also p. 79). It is by reputation particularly graceful and, to judge by its relatives, this might be a good thing in the garden. It is a monocarpic species with an evergreen winter rosette and a flowering spike which presumably expands after two to three years of growing to a dwarfish 45 cm (18 in) with one to three flowers on each panicle. The blooms are about 3 cm (1½ in) across and composed of 4 pale yellow petals. It requires standard growing techniques but, with something newly re-introduced, it might be safer with winter rosette protection.

M. LYRATA

A small and rather insignificant monocarpic plant that has been transiently in cultivation.

This is a widespread species from Sikkim to China. It was placed by Taylor in a different series to many of the other winter-dormant monocarpic species, such as *M. horridula*, for a number of botanical characteristics, but it would clearly resemble a rather poor form of some of them and indeed Taylor suggests the similarity with a form of *M. lancifolia*. The growth is weak and the flowers are few in number with sometimes only one per plant; it has been described as an annual although I doubt if it is. The leaves are highly variable over the wide geographical range and they may be round, elongate or lobed. The basal leaves die as soon as the flowering spike starts to expand which produces an ugly plant not unlike some of the worst forms of *M. horridula*. There are reported pink and white forms as well as the consistent pale blue, but clearly the plant has little to offer the gardener given the magnificence of related species with the same growth characteristics. The one really interesting characteristic, only shared with *M. chelonidifolia*, is the ability to produce vegetative buds in the leaf axils which presumably could be rooted on to produce new plants. Seed has recently been collected but no plants have been reported as flowering but it is likely to turn up again. Clearly it is a plant to be tried only by specialists in screes, unless the vegetative reproductive characteristics can be bred into better plants.

M. NAPAULENSIS (Plates 13 and 14)

The easiest and most available monocarpic rosette-forming species, highly variable in cultivation. One must immediately add that there is only one form (the blue-flowered one, called 'Wallich's Form') that is the true species, all the rest are at least partially hybrid particularly with *M. regia* and *M. paniculata*.

This species is easy in semi-woodland situations with good rich feeding. It can however be quite happy in a well-composted herbaceous border and will put up with a great deal of sun in more northern regions. The plants take between two and five or six years to flower but three or four is normal. The plant is evergreen and gradually expands in size until a flowering spike is eventually produced in late spring. The leaves can be highly variable in colour from a silver-grey to a brilliant ginger and this colouring is largely due to the very substantial number of hairs on both sides of the leaf surface. There are great differences in the form of the hairs and even the angle at which they arise from the surface. The leaves are also highly variable in the degree of dissection. Some are almost smooth in outline with slight notching of the edges while others are deeply lobed giving an almost frilly appearance. In general the ultimate size of the rosette is determined by the richness of the feeding of a particular plant but there is also great variability in potential.

The flowering spike can vary between 90 cm (2½ ft) to over 2 m (6 ft). Feeding is a factor in ultimate height but some strains are genetically much more compact than others. The flowers are borne on stems (technically cymes) that arise from axils of the upper stem leaves. The flowers vary from white through blue to yellow, pink and deepest ruby-red with every muddy colour in between. They can be multi-coloured with mixes of red and yellow or a properly mixed combination of the two in a delightful apricot colour. The flowers can be flat, cup-shaped or a perfect goblet-shape like a tulip. Some plants can have up to ten flowers on each flowering stem, though Taylor describes up to 17 arising from the axils of the lower stem leaves. The majority of plants in cultivation have between three and five flowers per stem and very poor forms may only have one. In good forms the whole flowering stem is very compact with up to 50 or more flowers at once. Flower size also varies and the yellow forms are usually largest at 8–10 cm (3–4 in) though this can be rivalled by some pink varieties. In general the red and the blue forms are smaller.

It must be realised that what we now grow in our gardens is not *M. napaulensis* but a hybrid. It has in the past interbred freely with *M. paniculata* and with *M. regia* and these hybrids are as fertile as the true

19. Flower panicle of *M. napaulensis*; the number of buds may vary between one at the top and a dozen or more lower down on good forms

species. It can also hybridise with *M. dhwojii* but the progeny of this cross are sterile. The large smooth leaf and large yellow flowers are derived from *M. regia*, the dissected leaves and small yellow flowers from *M. paniculata*. There is however great variation in *M. napaulensis* itself.

There was also another species, *M. wallichii*, which had two forms: var. *fusco-purpurata* with purple-red flowers and var. *typica* with sky-blue flowers. These two species were firmly put into the species *M. napaulensis* by Taylor. Since that time, particularly after the Stainton, Sykes and Williams Expedition, many more coloured forms of *M. napaulensis* have been brought into cultivation as well as red *M. regia* and a new pink species, *M. taylorii*, and all these are now well and truly mixed. The wreckage is not however unattractive and something can be done to sort out good forms and isolate true breeding strains. Many of the SSW collections were very distinctive and to this day SSW numbers are offered in seed exchanges, though this is quite inappropriate after so many generations of casual seed collecting from mixed populations in the garden. Nevertheless throw-backs that resemble the originals do occur and the best example is a form with the most perfect large china pink flowers and foliage that is very densely covered with brightest ginger hairs. It is possible by cross-pollinating between good and similar forms to select strains that are true to colour and approximately true to form. The red forms and the large yellow forms will also breed true. Selecting good forms of meconopsis has been dealt with on p. 35 but in no species is it more appropriate than this one.

20. Comparisons of Wallich's form of *M. napaulensis* (left) and the more usual form (right); they are distinct even at the same seedling stage

The blue form of *M. napaulensis* is sometimes called *M. wallichii* and sometimes called 'Wallich's Form'. It clearly is *M. wallichii* forma *typica* before Taylor lumped it into *M. napaulensis*. It has two very distinctive characteristics other than the beautiful sky-blue colour (again not affected by acidity of soil like the *M. grandis* forms). It flowers almost a month later than normal forms of *M. napaulensis*, still producing flowers well into September. The highly dissected yellow-green foliage is very distinctive, except possibly from *M. paniculata*. This form also tends to flower as a biennial and as a somewhat smaller plant. This is neither the time nor the place to re-open debates about what is a species but it would certainly be a pity to lose this lovely form by mixing its blue blood with paler shades. It does interbreed with other *M. napaulensis* forms producing a muddy-pink crossed with pink forms, a muddy-purple with red forms and the standard, and in this case rather delightful, cream with yellow forms. These seedlings are almost invariably sterile (which further convinces me it is indeed a true species). The best way with a mixed population is only to harvest seed from the bottom flowers of this plant which will be in flower after everything else in the garden is a dying skeleton. Saving seed from these monocarpic forms is a bit of a burden because the dying, but seed ripening, plants are nothing if not untidy and if I did not love them so much I would call them downright ugly. It is necessary however only to save seed from very good forms, and unwanted plants are snapped off at the base to be turned into the most nutritious compost. The roots will have to come out in autumn when you remake the bed and the degree to which the vast fibrous root ball has drained the soil of goodness is a powerful incentive to feeding the next generation properly.

M. napaulensis are not in general difficult to grow using standard conditions and usually self-seed, which is not a good idea since in the long run it will produce a vigorous but coarse strain. At least 45 cm (18 in) should be left between plants when bedded out. They are generally fairly resistant to whatever winter can do to them with frost, snow or rain but would probably succumb to freezing and thawing in a saturated bed. I have tried these plants from wettest, coldest Wales to driest east of Scotland and the losses each winter are insignificant. I know in some areas they do need covering against winter wet if it is persistent, but this is something that has to be learnt for each area. It is essential if you do cover them that the ground underneath the main root ball does not dry out completely. Covering spoils a major attraction, with their diversity of colour and form in the drab months of winter: indeed, it is my ambition to persuade a parks department to plant a bed of them as a winter feature—they are also very suppressive of weeds so I think it is a potential to develop. A few spare plants in pots could always be used to fill the space of the odd accidental loss.

Although it goes against the number one dictum—meconopsis hate pots—it is possible to overwinter in pots and plant out later. Later can even be two years later since a friend had my stunted plants in pots when my own were 1 m (3 ft) across and flowering. (I would not recommend this but it is an indication of their toughness.) They are pretty resistant to summer drought especially in a shaded situation and would probably put up with a great deal if they were planted over a really deep layer (up to 30 cm, 1 ft) of compost or manure. They are quite happy in the herbaceous border as long as not overgrown and will respond to a poor soil if fed regularly with a good balanced inorganic fertiliser.

They are widespread in the wild; both high steep grassy ridges as well as river banks have been described as habitats. It is also a plant of woodland meadows and open clearings, and has been described in shaded but rocky gullies. Clearly the variable habitats are indicative of a plant that has an innate tolerance to many different environmental variables. There is evidence that some forms are better and easier than others and the plants that were brought into cultivation by the 1954 Stainton Sykes and Williams Expedition have made a great impact.

M. NEGLECTA

The western-most Himalayan species from Chitral but possibly only a form of *M. aculeata*.

This was a species described by Taylor from west of the river Indus and is the only Himalayan species that occurs there. Taylor suggests that there may be other species from even further west but there have in fact been no further reports in over half a century. This species is clearly closely related to *M. aculeata*, having blue flowers. It is distinguished by the flowering stems all being basal and by having a particularly short style. Taylor himself pointed out that neither of these characteristics are reliable and are well known to vary in other species. The basal flowering characteristic in particular occurs in *M. aculeata* itself. This population is however geographically isolated and if it is not a true species it is clearly well on the way to being one. All this was based on a single collection of a single specimen, but whatever the status of this plant its position at the dry end of the Himalayas makes it interesting.

M. OLIVERANA

A species of botanical interest only, not in cultivation but significant in the evolution from *M. cambrica*.

This species belongs to the small group of Himalayan plants that are polycarpic and have yellow flowers and show some resemblance to *M. cambrica*. This species has never been in cultivation and comes from eastern Szechuan. It has a bristle-covered upper rootstock which is characteristic of this species and its relatives, *M. chelonidifolia*, *M. villosa* and *M. smithiana*. It has an erect and branched flowering stem with many of the lower leaves dead at flowering time. The leaves themselves are divided and the plant is obviously very similar to *M. chelonidifolia*. The major difference is that the seed capsules which are long and narrow splitting for only a short distance are more similar to *M. villosa*, even though Taylor placed this latter in a separate series to *M. oliverana*. *M. chelonidifolia* is a nice delicate and well-behaved wildling for the woodland garden inclined to be taken advantage of by other plants and gardeners, the latter usually by neglecting it! *M. oliverana*, if it were in cultivation, would clearly need the same treatment though whether two poor relations at the same time would appeal to anyone but the most ardent species collector is doubtful!

M. PANICULATA

A robust monocarpic evergreen rosette species with yellow flowers—and a reputation for being easy.

To most gardeners this species looks like a yellow *M. napaulensis* and indeed in many cases forms a part of the latter's hybrid swarm. There are two variations (un-named) of the plant in cultivation at present. One from Bhutan has grey leaves and the other from Sikkim a lime-green but the flowering spikes of both are identical. The taxonomic characteristic that distinguishes this from yellow *M. napaulensis* (and from *M. robusta* for that matter) is the substellate pubescence of the hairs on the leaves. This is fine in

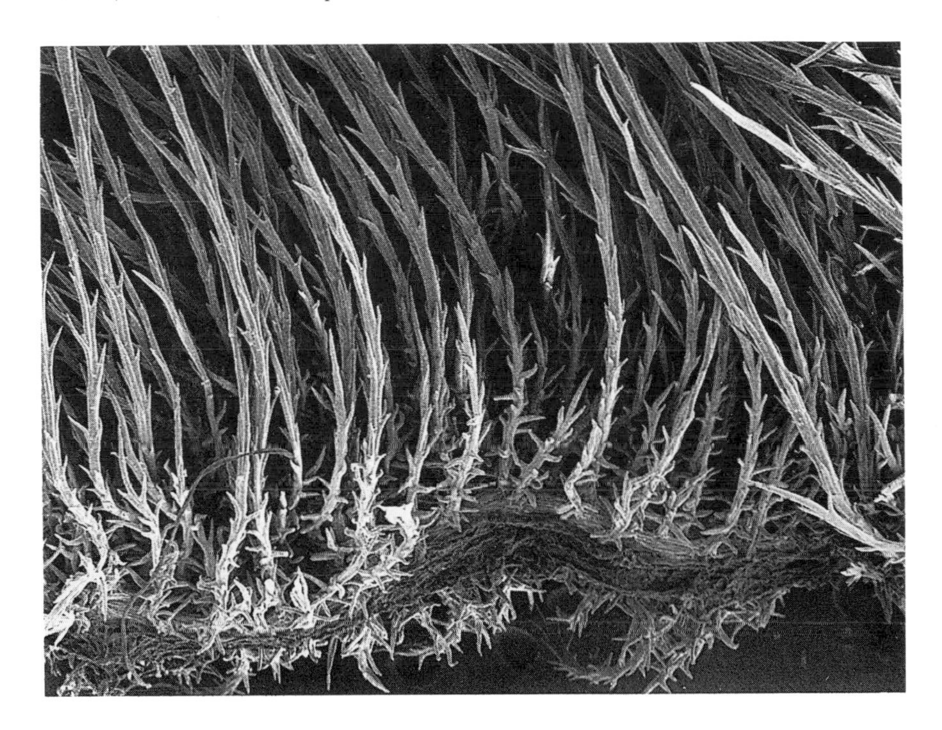

21. Comparison of substellate pubescence of *M. paniculata* (above) with *M. napaulensis* (overleaf), this character is critical to identifying the former species from similar yellow-flowered species and is difficult to distinguish even with a scanning electron microscope

theory but in practice can be a little trying even with a microscope (Figure 21). The leaves of all these species are variably covered in hairs and of course many *M. napaulensis* are crosses of *M. paniculata* anyway so comparison is difficult. There is no doubt that the true *M. paniculata* is in cultivation from a number of different and geographically separated places. There is fortunately another characteristic that is reliable—the stigma is a soft purple colour.

Examination of many hundreds of *M. napaulensis* hybrids from all sorts of places and with all colours never shows them to have anything but a green stigma. Many of these green stigma forms of *M. napaulensis* have *M. paniculata*

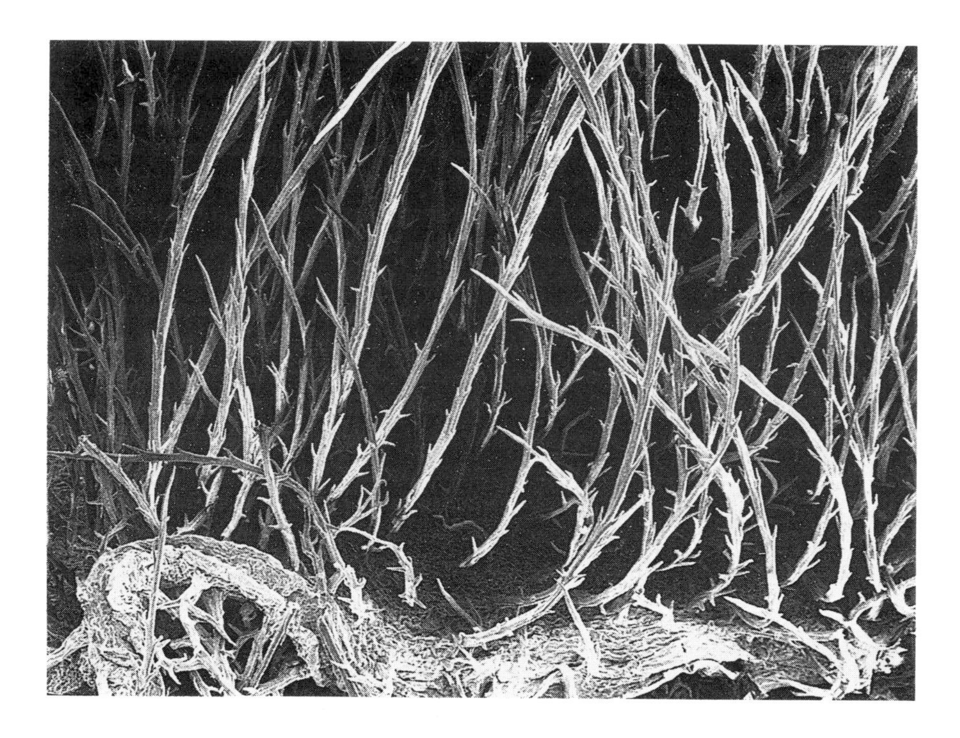

21B.

blood in them, so to speak, and clearly the purple colour is recessive. The same purple stigma (though this time a deeper colour) separates *M. regia* from *M. napaulensis* and this is also suppressed in the undoubted *M. regia* × *M. napaulensis* hybrids that abound ('× regia'), all having green stigmas.

M. paniculata is easy from seed and should be grown using standard techniques. Like many of the evergreen rosette-forming species the plants remain rather upright into their first winter and the flat rosette shape gradually develops in the middle of the dying leaves of the first year. *M. paniculata* tends to flower at two and three years old as opposed to three and four years old with *M. napaulensis*. Really good rich growing conditions will favour flowering at two years. The flower spikes emerge in late spring and are not normally vulnerable to frost. Neat cup-shaped flowers between 5–7 cm (2–2½ in) across open from the top and are of a nice clear yellow. The flowering shoot is normally about 1.5 m (4–5 ft) tall, though occasionally up to 2 m (6 ft), and there are usually single flowers on each stem arising

16. M. quintuplinervia

17. *M. sherriffii*

18. *M. superba*

19. *M.* × *finlayorum*

20. *M.* × *sheldonii*

from the upper axils of the stem leaves but lower down there may be up to 4 or 5 on each pedicel. All variations in between occur however and Taylor found no justification in separating them into any sort of botanical rank.

This species has been described as much tougher than *M. napaulensis* and that it does not suffer from winter crown rot and can withstand drier conditions. It undoubtedly is the best species to start with in areas that are marginal either because of the dryness of the summers or the nastiness of the winters (in particular wet and cold in random and mixed amounts!). This toughness has been disputed elsewhere and there can be two reasons for this. It may be that seed from different areas has different characteristics. Or that plants respond quite differently in apparently similar places and climates and merely reflect our continuing ignorance of many of the factors that our plants have to cope with. It is quite likely that it is both these things. There is no doubt that good rich feeding, a modest amount of shade, an open textured soil and no winter waterlogging are what are required. A slightly sloping site is always better for all these rosette meconopsis: raking things level only leaves surplus water lying in puddles around our plants. A glass tent cloche in winter cannot do any harm as long as it merely keeps off most of the rain and lets the wind blow through, but *M. paniculata* is probably the species that needs this least.

The species is very widespread from Nepal to Assam and occurs in high alpine meadows and boulder screes as well as at much lower levels in scrub at the edge of coniferous woodland. This widespread occurrence is no doubt the reason for the adaptability in cultivation and suggests that even in very marginal areas that strains could be developed that would survive, if not flourish.

In cultivation unless kept quite separate from related species it will hybridise. The movement to form national collections of plants in Britain is a most worthy cause but some new thought must be given as to how to maintain monocarpic species that are inter-fertile. I would personally arrange a register of people prepared to grow only one species and maintain purity in that way.

M. PINNATIFOLIA

A newly described species with purple flowers from Tibet related to *M. discigera*.

This is another species recently described by the Chinese. It appears to differ from *M. discigera* by having dissected leaves and was described from Jilong in Tibet near the border with Nepal. It has been illustrated in colour in *Chinese Alpine Flowers* (1982) and although some species of meconopsis in that book are incorrectly identified, this illustration shows a plant closely related to *M. discigera*. It must be borne in mind however that plants of *M. discigera* itself were described by Ludlow and Sherriff that had dissected leaves. One can say no more without access to the relevant herbarium material and this is in China! It would presumably be similarly difficult in cultivation to *M. discigera*.

M. PRIMULINA

A rare deciduous monocarpic species with blue flowers, not in cultivation.

Taylor placed this species along with *M. lyrata* in a series of their own. They are small winter-dormant species that would superficially resemble *M. horridula* and its relatives. There are a few flowers in the axils of the upper leaves. The plant is distinguished by the flowers being borne singly on long pedicels that arise from near the base of the flowering stem. There are only relatively few such flowers and the whole plant appears insubstantial. It is recorded from the borders of Bhutan and nearby Tibet at altitudes between 4,000 and 4,600 m (13,000–15,000 ft).

M. PSEUDOVENUSTA

A purple-flowered relative of *M. horridula* and thus deciduous and monocarpic.

It is found in south-east Tibet and across the border into China at 3,500–4,300 m (12,000–14,000 ft). Taylor split this species from the similar *M. venusta* primarily for the differently shaped fruiting capsule. In this species it is shorter and broader than in *M. venustra*. The flowers in *M. pseudovenustra* are often more than 4-petalled which further characterises it. Both species are close to *M. impedita* and all have the general growth characteristics and form as *M. horridula*. It is probably a desirable garden plant and certainly one of the group related to *M. horridula* that has desirable purple flowers. There are many persistent leaf bases present in plants in the wild and the examination of these suggested to Taylor that this plant would take several years to flower rather than being a biennial. It is found in stony pasture away from trees or scrub and this suggests that a similar habitat to *M. horridula* would be satisfactory, with a moderately moist summer and a dryish winter with good drainage. It is interesting that none of the purple relatives of *M. horridula* has persisted in cultivation since seed has been sent home of a number of them, such as *M. lancifolia*, *M. impedita* and probably others. It is possible that seed viability is a problem but it is more likely that these species from the Chinese end of the Himalayas need a modification of growing techniques. A warm dry spring before the onset of the monsoons may be a significant feature.

M. PUNICEA (Plate 15)

A blood-red gem from China of exceptional purity of colour, it is a difficult plant that is doubtfully perennial.

This species until recently was a fabled memory in gardens from more than 25 years ago. It was a plant that Farrer raved about, the long red petals reminding him of flags streaming in the wind and now that it is back with us it is possible to understand his feelings. It is a species from Tibet and China reaching north to Kansu. It has been introduced to cultivation a number of times and lost, but such is its brilliance every effort to keep it is required. It resembles in habit *M. quintuplinervia* though it forms very tight clumps and is more substantial in leaf and flower. The seed from the Cox and Hutchinson expedition in 1986 produced two types of plants (C.H. and M. 2586). One has ginger hairs on the leaves and a more open rosette-like form when young; the other has less prominent pale hairs making the leaves look plain green and the young plants are more upright and delicate looking. They are difficult to tell apart when mature but the seed pods of the ginger-haired form are slightly rounder.

This species makes substantial sized plants by the second year, if well-fed, and has a complex mass of basal leaves appearing to make up a large number of sub-crowns. In a large plant 30 cm (1 ft) across there may be up to 20 separate flowering crowns and by the end of June a plant can have over 50 flowers in its second year. The flowering stems are all basal and elongate over a number of weeks—they may be over 60 cm (2 ft) or more by the time the flower expands—with the flower bud gradually increasing in size. The opening of the bud is one of the sensations of the plant world especially when seeing it for the first time. The petals are five or six times as long as the bud and emerge as long pendent streamers of the most perfect red. Even the best blue poppies take a few hours to develop perfect colour but *M. punicea* is a glowing fire at once. The packing of the petals in the bud is a minor miracle and they are charmingly creased as they emerge and need warm sun to iron them smooth. Such is their glamour that to criticise the creases in their dress would be ungallant. They are long-lived as flowers, in full bloom for a week and they dance and swirl in the merest breath of breeze and stream like pennants in the wind. The flowers are fully pendent and tightly closed in the cool of the day but open slightly in the warm to let insects in. The pollen is a dull purple-brown and not abundant. After petal fall the capsules gradually turn fully upwards. The flowering period is months and were not the seed so precious they could easily be dead-headed. They are likely to flower in bursts as do good forms of *M. quintuplinervia* and there will be odd flowers right into autumn. Indeed this is probably the whole reason for the difficulty of the plant since it does not know when to stop and will flower right into winter, instead of going dormant, and then die of exhaustion. There may be forms that are more perennial than others as is the case with others of the polycarpic species and our long-term success may require us to select for them. The seed will of course be produced over the same long time span as the flowers, taking about four weeks to ripen. The seed stays in the capsules reasonably well because they are upright when mature but should be harvested on a continual basis.

After only two years with it back in cultivation one can see three reasons why the plant has been lost before. The seed is difficult to germinate, the fertility of the plants is low and, although the plants give every indication of being perennial from above ground, with apparently many separate crowns, they all radiate from a single tap root; if this dies after flowering (and it apparently nearly always does), all dies and division is not possible.

It is essential to sow the seed in the autumn and overwinter it ungerminated in a pan. The seeds are large by meconopsis standards and should be sown evenly and thinly. Spring-sown seed fails for nearly everybody. The wise who keep their seed pans till the following spring are rewarded with germination then. Nevertheless, this is not worth the risk and seed must be autumn-sown if possible. This peculiar behaviour of dormancy is not really understood and until it is, and we know how to break it, we must put up with the phenomenon and do the best we can. They need standard treatment from seed and are not difficult. If they are really well-fed a few of these plants produce flowers by early October but they are poor shadows of the brilliance to come in spring. As already mentioned, flowering of this species in winter is a problem that has been noted before during earlier periods when it was cultivated. They have a tremendous propensity to flower and a reluctance to become dormant. They are clearly a very early flowered species and two plants in a cold alpine house with no protection were never fully dormant all winter, gradually growing larger and more robust even in January. The buds were visible early in February and may well have been formed in the autumn. The one encouraging feature is that they appear quite frost hardy. They survived –8°C (18°F) in flower when *M. sherriffii*, covered and ostensibly fully dormant, once again lost all its buds. Friends who grew the same seed in less vigorous condition overwintered much smaller plants, some of which will not flower until the third year. The plant is clearly not very perennial and keeping half the plants back in this way is probably very desirable and should be a routine cultivation strategy.

It seems that only exceptionally will any meconopsis set seed if self-pollinated and it is clearly advisable to have a whole group of plants together, and with something as rare and choice as this species, to hand pollinate. Careful hand pollinating has made it clear that there is not an abundance of pollen and that on cold or cloudy days the flower would present a pretty formidable obstacle to any pollinating insect, at least in the UK. I suspect a bumble bee by sheer bumbling would find its way in somehow. One must assume that the pendent nature of the flowers is not by chance and at least one reason would be to keep what little pollen there is dry. There is good seed in some pods but not in others and it may well be that this species needs a really warm spell to produce seed. The lack of seed has been noted in the past and a poor year with no seed might once again set this species on the road to extinction. It might be necessary to keep a few plants in pots in a warm greenhouse if necessary and hand pollinate. It is never clear with phenomena such as this low fertility whether it is a consequence of cultivation conditions, or whether they are adapted to a different regime of pollination in the wild where a very modest investment in pollen is associated with a rather specialised pollinator.

The problem of the non-perennial nature in many ways is the most worrying. They have not so far been difficult, far from it, they have been remarkably accommodating and grow happily in full sun and shade. Species like *M. grandis* and *M. betonicifolia* when grown from seed produce a variety of plants some of which are much more likely to be perennial than others. The ones that are soundly perennial are those that produce several crowns but also and absolutely vitally, a separate root system for each crown. In this state the plant can simply be divided carefully between crowns and separate plants grown on. Both these species produce side buds from the root stocks that can also be detached and rooted like cuttings. *M. quintuplinervia* is the most perennial in this sense and can go on being divided almost indefinitely since almost every part of the plant has separate roots. *M. sherriffii* on the other hand appears to be like *M. punicea* with a single rootstock however many crowns above ground and thus cannot be safely divided. It might just be possible to slice through the whole plant and regrow them on but the risk of fungal infection would be enormous. One can only hope that some plants will produce a multiple rootstock and that vegetative propagation can be achieved from these forms. It may be that a small proportion of *M. punicea* and *M. sherriffii* do have this characteristic and there is some evidence that this is so for the latter. Plants should be kept covered in winter with bracken or conifer branches placed over the crowns if really severe weather, particularly bitingly cold winds, occur without snow cover.

In the wild it grows in damp meadows among rhododendron scrub and is at least partially shaded.

M. QUINTUPLINERVIA (Plate 16)

Farrer's 'Harebell Poppy' and a gem of a perennial deciduous species with soft pale lavender flowers from basal scapes.

This is a lovely species of very perennial nature which is at home in a rock garden setting as well as a peat bed. It is easily propagated vegetatively which is just as well as it is especially difficult from seed. It is in the same series as *M. simplicifolia* and *M. punicea* but is also a close relative of the *M. grandis* series of species. It has a fibrous root system and the whole plant is bristly (as opposed to spiny). The leaves are simple and emerge from basal rosettes. They can be up to 25 cm (10 in) long, part of which is petiole. The leaf rosettes emerge in spring from full dormancy below ground level, but much of the dead leaf material remains above ground throughout the winter to mark its resting place. Care is needed overwinter since unduly zealous tidying up and pulling away dead remains can damage dormant but living rosettes. The flowers emerge from basal scapes and usually only one from each rosette of the plant. The flowering stem extends between 15 cm (6 in) in early flowers on compact strains of the species and up to 60 cm (2 ft) on late flowers on more robust plants. There is no tendency to flower themselves to death although they will often throw a few late flowers right through the summer into September. The flowers in general are pendulous or semi-pendulous and are a neat cup shape about 3 cm (1½ in) in depth and breadth. They are generally a pale lavender of great purity of colour. The anthers are an attractive cream to buff colour and enhance the quality of the bloom. The whole plant is one of those rare plants at blooming where the perfect proportions of all parts immediately set it aside as one of the exceptional plants in the garden world. The fame of the genus is undoubtedly the classic blue poppies but this one is a real flawless gem. Farrer gave it the name the 'Harebell Poppy' (which to me at least has lovely romantic rural connotations) particularly because it had such an awful specific name given to it by Regel.

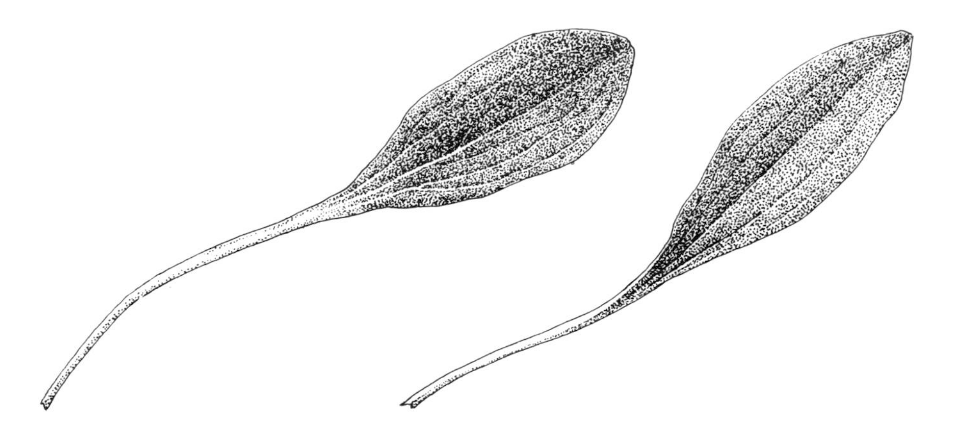

22. *M. quintuplinervia* leaves

There are problems with the large seed but the reason is not clear. Some strains merely produce dust which is clearly not viable. There are other strains that produce what is apparently good seed. This will sometimes germinate the following spring, if sown soon after harvesting, but this varies from year to year. It will almost never germinate if sown in spring which means problems for people trying to establish this species from seed obtained from various seed exchanges. In my own garden I have produced most plants by carefully hand weeding around parent plants and leaving meconopsis seedlings until one can identify them to species. This is not difficult since by six weeks the narrow true leaves, like delicate elongate editions of those of *M. betonicifolia*, distinguish them.

The behaviour of meconopsis seed and its strange dormancy characteristics have been discussed earlier. If you are dependent on seed exchanges then ask the secretary to put you in touch with someone who will send seed in the autumn. (The hard-working people who run the exchanges are well aware that dormancy for some species may be unbreakable at present after winter drying and would be quite happy to see such seed going to a good home as long as they do not have to become involved in the distribution!) Meconopsis are generally self-infertile and it would be worth planting two different clones close together to see if the viability can be increased. It is possible, though it does not appear to be the case, that the species has been vegetatively propagated for so long that it is genetically flawed. It also gives one confidence that meconopsis may be relatively unsusceptible to virus infection; plants vegetatively propagated for so long would surely show symptoms after half a century.

Propagation by vegetative means is easy and less risky in this species than any other I have handled. It is always best done in spring when plants can be grown back to strength, although I have broken a plant up at any time of year including mid-winter as long as the ground was not frozen. The reason for this confidence is twofold. First, it has a remarkably tough constitution; second, the mass of fibrous roots sprouts from just about everywhere below ground level and almost all pieces will have some root if pulled off and will root like cuttings if they have not. One word of warning here is that none of the species will stand the saturated polybag over flower pot technique. A really gritty compost with some rooting compound (as much for the fungicide usually mixed in, as the hormone) and as open a cutting frame as is compatible with not dehydrating them. It is always safer to do a simple division in half, by gently pulling at a dug up 15 cm (6 in) clump, than trying to cut off with a knife a more mean piece. This plant takes up little room and four or five plants can rapidly be built up and a couple of these can then be kept for making noble and generous gestures on the spot to distinguished visitors. There is in fact a patent safety way of propagating this plant which has been mentioned earlier under *M. betonicifolia* and involves surrounding the plant in spring by a moat of very rich compost (see p. 55). By late summer several new rosettes rooted into the special mix can be pulled or cut off with no risk to the parent plant.

One might wonder why more gardens are not carpeted with this plant, especially since seed is so unimportant that the dead head can be cut off to stimulate a second or even third burst of flowering. It must be confessed that it is probably a little fussy about where it will grow happily and though it is not subject to sudden death without warning, like so many of its relatives, it may just fade away. It would grow well in a rockery but not a hot scree or a dried out pocket in the 'dogs graveyard' type, with large lumps of rock placed in regular rows. A nice deep area of rich leafmould would be ideal and the further south the garden, the more northerly the exposure. It will certainly take full sun in the north and is all the better for it. It becomes lank and straggly in a woodland setting with deep shade but produces lovely large floriferous plants in a sunny clearing. It is certainly a limestone plant but

I suspect it is simply not fussy as to pH and will take what you give it; but, like all meconopsis, it does expect decent rations and probably responds to inorganic feeding if that is all that is available. Its needs are so modest compared to its robust relatives with a larger appetite that some modest amount of compost could be found to fork in around it and it should be split up and replanted in fresh soil every three or four years.

There are at least two forms in cultivation. One is larger and more straggly which sets less seed and the offshoots from the main plant produce a rather untidy growth after a year or two. This form is much more vulnerable to disturbance by cats and blackbirds and is definitely inferior to the other form. This came to me as 'Kaye's Compact Form' and we have much to be grateful for to the distinguished British nurseryman, Reginald Kaye from Silverdale in Lancashire, who found it in a local garden but there is no evidence as to exactly where it originated. This form gradually expands until after about three years it is a compact mass of neat green bunny-eared leaves from which up to 50 uniformly coloured pale lavender flowers will emerge over about six weeks from mid-May onwards. The plant invariably stays compact but its seedlings do not come true. This plant undoubtedly sets some viable seed and is thus self-fertile.

The faint worry about winter hardiness relates to persistent reports over the years of losses in exceptionally hard winters. The plant is very near the surface at all times in winter and, although the buds are fully dormant in the dead remains of last year, bitter cold winds could erode the soil and desiccate the whole root system. A slightly raised bed and a simple glass tent cloche with dried leaves tucked in either deliberately or by the wind should eliminate the risk. The effort of doing this is minimal and since the plant is so utterly obliging in all other respects this cannot be denied it. I have to confess I have never offered it any protection and it has never yet needed it. It is late into growth in spring and there is no evidence of spring bud damage from late frost. It has superb potential to the hybridiser and I foresee a collection of really excellent rock garden plants being derived in the fullness of time. There are already the hybrids with *M. punicea* and *M. integrifolia* (*M.* × *cookei* and *M.* × *finlayson* respectively) and Margaret and Henry Taylor of Invergowrie have recently produced another excellent un-named hybrid with *M. betonicifolia.*

It is found in the wild in more northern areas of western China and Tibet where the influence of the monsoons would be less significant and at a range of altitudes from 3,000–6,000 m (7,000–14,000 ft). This is effectively from temperate meadows to harsh alpine conditions. There is some variability in the wild with darker blue-purple forms and albinos and although deeper blue ones are recorded in cultivation those of white ones have been transient though they would be eminently desirable.

M. REGIA

An aptly named monocarpic evergreen species with large yellow or occasionally red flowers, AM 1931.

This species is regularly distributed and even available commercially but there is no doubt that all on offer are hybrids. This species typifies the winter-rosetted species and is distinguished by two features, both necessary for correct identification. The leaves are unlobed with a narrow saw-edge to them and of very substantial size. The colour of the stigma is a deep black-purple. There are a number of illustrations in various textbooks of this species and both characters are obvious. There are also a number of other illustrations in what purport to be definitive textbooks that are of the hybrid swarm. It is likely that both *M. napaulensis* and *M. paniculata* are involved in the hybrids but in the progeny the leaf is always at least faintly lobed and the stigma invariably green. In many of the hybrids the colour and size of the flowers are excellent and no way inferior to the true species; it is only to botanists that the differences are anything but trivial. The hybrids are fully fertile and if new seed of the true species is obtained from the wild then they will have to be grown apart from all similar species.

23. *M. regia* leaf

True *M. regia* forms an evergreen rosette which can reach over 1.2 m (4 ft) across before it finally sends up a flowering spike. The whole plant is covered by thick golden hairs giving it a characteristic glow, especially in winter with frost sparkling upon it. The leaves are substantial, and can individually reach at least 60 cm (2 ft), with a short broad leaf stem, and are up to 10 cm (4 in) across. The outer leaves die off as the plant ages and new ones are produced continuously from the centre. The plants, because of this, slightly contract in size each winter as the largest outer leaves die away.

The flowering spike emerges normally at three or four years old. A few may wait an extra year but any flowering at two years old are likely to have an undue influence from the more precocious *M. paniculata*. The flowering stem may reach up to 2 m (6 ft) in rich conditions with a moderate amount of shade to draw the plant up. The flowers are of a soft yellow only rarely as intense as the best *M. integrifolia* and these can be up to 13 cm (5 in) across. Plants with flowers much smaller than this are likely to be closer to one of the other species, but there is a complete range of size. Densely hairy flowering stems emerging from the axils of the stem leaves and at the top of the stem may only have from one to three flowers on each of these panicles, but lower down they can have up to ten or more flowers in exceptional forms and are thus floristically very spectacular. As with all their evergreen rosette relatives they are magnificent in leaf, stately in flower and pretty foul in seed. It is possible in a reliable climate with a good autumn to cut off the top metre (3 ft) after flowering and save seed from the lower flower capsules that are less obvious for the rest of the summer. A few seed pods will produce an abundance of seed. The plants will invariably die after flowering.

Red forms of this species with the necessary leaf character and purple stigma colour were introduced by the Stainton, Sykes and Williams Expedition and there are good illustrations of this plant, but it too has become submerged in the hybrid swarm. Taylor described two types of the yellow form, one with golden hairs throughout its life and another with paler silkier hairs that only became golden on maturity. The latter was a more robust growing strain.

Three generations of hand pollinating have produced a large flowered, highly floriferous and shapely replica of *M. regia* but the stigma is still green and leaves slightly lobed. It would fool all but the expert. The plant that is actually most closely related to *M. regia* is *M. superba* and Taylor put these two in their own series separate from *M. napaulensis* and others. *M. superba* is instantly recognised by its silver hairs and brilliant white flowers but is almost identical otherwise. Williams described a pink-flowered species with a dark brown stigma as *M. taylorii* that is also similar to *M. regia* in general form. The nearest thing I have seen to true *M. regia* was an undoubted, and as far as I am concerned unique, hybrid between *M. superba* and *M. paniculata*. The flowers were of a cream-yellow but the rest was textbook. The hybrid plant was sterile which suggests that *M. superba*, one of the parents, is genetically more isolated than *M. regia* from its relatives. I suggest *M. paniculata* as the other parent because this too has a pale purple stigma, purple appears to be recessive to green, and thus both parents needed to have purple stigmas to produce a hybrid offspring with a purple stigma.

Seed should again be sown thinly and plants grown on in standard conditions. As always they should be planted out as soon as possible but I have always kept a few spare plants in pots overwinter to make up for the occasional winter loss. In large trial plantings of many types of rosette-forming species in a totally open and unprotected bed, only forms that closely resemble *M. regia* were absolutely unharmed over three or four winters. Although I have never grown it, there is no reason to suppose that this species is vulnerable to late frosts. A few of the hybrids may show contorted foliage in spring that looks virus-like but it is the effect of cold. It is not clear whether in a mixed batch of seedlings containing all manner of flower colour and leaf shape whether this damage is random, and related to subtle differences in the micro-climate surrounding different plants, or whether there is a genetic disposition to lack of spring hardiness and this is gradually being selected out. The *M. regia*-type hybrids are the least susceptible to this problem and some of the small red-flowered *M. napaulensis* the most susceptible.

These are desperately greedy plants and I use the word advisedly. The soil occupied underneath these plants, which may be half a cubic metre (yard) of soil, has to be seen to be believed when you dig out the dead remains. One finds a tap root as thick as one's arm with a plunging and spreading mass of fibrous roots. The soil is chalky looking and dry with little sign of humus. I have found if more than two generations of these plants are grown in the same site they do not prosper as well as plants in new areas, however much enriching material has been dug in before replanting.

The species is found in the wild in central Nepal and was not brought into cultivation until the 1930s. There were re-introductions in the 1950s from expeditions to Nepal. These expeditions questioned the relationship between *M. regia* and forms of *M. napaulensis* and the whole question of hybrids in the wild. There is no real evidence that *M. regia* occupies a different habitat from *M. napaulensis* and it occurs in the same geographical locality. Nevertheless, examination of herbarium material and photographic evidence makes it pretty clear that they are truly separate species and intermediates between them in the wild do not represent a gradation between extremes but hybrids. One must assume that *M. regia*, like *M. superba*, became isolated from the common ancestor of the rosette-forming species but that recently the isolation barriers have disappeared and the two species are growing together again.

M. ROBUSTA

A poor monocarpic evergreen relative of *M. napaulensis* with yellow flowers.

Taylor only described this species from fragmentary material. It is the most western of the evergreen rosette-forming species and seeds of this species have recently been offered regularly by one of the Indian seed houses that collects from the wild. It is described from Kumaun at relatively low altitudes of 2,400–3,000 m (8,000–10,000 ft). In cultivation it has germinated without difficulty. It has resemblance to three species: *M. napaulensis*,

M. paniculata and *M. gracilipes*. It forms a modest sized rosette about 30–45 cm (1–1½ ft) across with dissected leaves and usually flowers as a biennial. The flowering spike is 1.2–1.5 m (4–5 ft) high and the flowers are borne singly on stems from the upper leaf axils and not more than two or three from lower ones. The flowers are the same colour as *M. paniculata* and usually not more than 2.5 cm (1 in) in length or breadth. It lacks the robustness and flower size of yellow *M. napaulensis*, for all its name, it lacks the hairiness of *M. paniculata* and it lacks the grace and fern-like degree of leaf dissection of *M. gracilipes*. The odd specimen could be confused with all three. The only ones that I have grown were put into a rather dry bed because of the western habitat of the species in the Himalayas and all could be described as pretty weedy and decidedly of botanical interest only. One must confess however that though they fit the description given by Taylor, he had an incomplete series to work from and it may well be a form of *M. napaulensis*. It has little value in the garden when compared to similar yellow-flowered species and is not in the same class as *M. regia*. It might however be valuable in drier climates since, although the winter rosette is not exceptional, it certainly is a reasonable substitute for the better species if these prove intractable. It might of course look a lot better if given rich feeding in a moist peat bed situation and is worth a try if you have the space.

M. SHERRIFFII (Plate 17)

A highly desirable pink relative of *M. integrifolia* that can be polycarpic, AM 1951.

This is a choice and undoubtedly difficult species, that has remained in cultivation since it was collected by Ludlow and Sherriff from Bhutan, due to the dedication and talent of a few growers, including for many years the Sherriffs themselves. There are still a small band of people who can grow this plant and they provide a small supply of seed and seedlings for others to keep experimenting with. The plant has simple strap-shaped leaves that resemble some forms of *M. simplicifolia* with fairly long leaf stalks. The seedlings always look rather delicate and the leaves at this stage are very vulnerable to even moderate winds. The plants may well flower in the second year from a single rosette on a tall flowering stem that has a few cauline leaves but usually only one flower. This is large and of a good clear pink with only rarely even a trace of muddy undertones. It closely resembles some types of *M. integrifolia* in form and shape. Perennial forms of the plant and seedlings in their first winter die right back but points of growth are still visible among the dead remains. If plants survive beyond the second summer they normally become multi-rosetted but seem to still be associated with a single rootstock which makes division difficult.

Seed is usually scarce even on good plants and much of what is offered is no more than aborted seed. The seeds are slightly tear-shaped and obviously plump when viable. They store over winter quite satisfactorily and usually germinate readily, though such is the scarcity of seed it is unnecessary to warn against sowing it too thickly! The seedlings grown on quite rapidly but are prone to damping off and care is required to keep seed pots moist not soggy. They should be pricked on at the usual stage and then potted on into 10 cm (4 in) pots. It is at this stage that life becomes difficult. They are very much a plant that insists on summer humidity. This is easily achieved in a shaded frame and as long as the plants are reasonably cared for in pots they will grow on. It would probably be quite possible to nurse them on to flowering in pots but they would almost certainly be monocarpic under such circumstances. Sooner or later they have to go out and as they are greedy feeders sooner is better than later. They do well in the type of misted scree described on p. 36. The constant watering rapidly leaches nutrients and a modest amount of resin-based slow release fertiliser for each plant, renewed in midsummer, is needed. The plants will then grow on most happily.

The plants require drying off in winter and small tent cloches with added dried leaf protection is good, though covering the whole scree as suggested on p. 36 is an alternative. The following spring the plant will be becoming multi-rosetted and should flower. Many will also flower even if they have not become multi-rosetted and this is invariably fatal even if the flower stem is removed. There is also a major frost hazard since the species is not frost hardy to flower bud damage even, probably, in mid-winter to the bud primordia, still in the very early stages of growth deep in the unexpanded crown of the plant. At this time it is of no great consequence since the plants can be kept well covered with dried dead vegetation, or even netting sacks full of polystyrene packing beads, but once into growth in spring they can only be briefly covered against overnight frost. The plant grows on perfectly normally but the top of a strong flowering stem is crowned not by an exciting and fattening bud but by a blackened carbuncle. At worst the plant will then die after 'flowering' and at best you will have to hope to do better next year.

The plants should be grown on after flowering, maintaining the humidity, until the critical period of induced dormancy. The species is susceptible to mildew if kept too dry in summer and is not robust enough to throw it off as is the equally susceptible *M. betonicifolia*. It is also prone to a crown rot in a damp autumn: this is how most of my plants have gone and the only cure I know for it is prevention by very skilled judgement as to when to dry them off a little and this includes making as accurate a guess as is possible about the long range weather forecast. They need covering in all subsequent winters. Certainly all the plants that I have examined dead or alive have only a single root system, however expansive they are above ground and however many rosettes there appear to be. This would make them impossible to divide. I have however been assured that this has been done. This allows a point to be made about variability. Some people, noted for being exceptionally skilled plantsmen, still appear to grow this plant with little difficulty, but it is not a plant that has been generally successful. Many individual plants of the polycarpic species, even *M. betonicifolia*, are probably monocarpic however they are treated. Finding a polycarpic plant needs patience since in some varieties only one in a hundred may be polycarpic. This may seem a lot of effort for a single plant but this is the joy of this genus, most are beautiful, some are easy but some provide great challenges. There is little doubt in the UK that the further south one goes the more difficult this species becomes and the more contrived the growing environment required, but at present this is the only good pink perennial species. A hybrid (in my garden but unnamed) of this with the monocarpic form of *M. simplicifolia* is robustly perennial and must give hope for a hybrid race of pink polycarpic rockery meconopsis.

M. SIMPLICIFOLIA

A species that at its best is unsurpassed among the true blue poppies but is often monocarpic.

24. *M. simplicifolia* 'Bailey's Form'

This is a relative of *M. grandis* though it is placed in the same series as *M. quintuplinervia*. It has a typical sky-blue flower but can be distinguished from species with similar flowers (notably *M. grandis* and *M. betonicifolia*) by having blue filaments to the pollen-bearing anthers. It also flowers from basal scapes, although there are forms of *M. grandis* that are like this. The leaves, as the name implies, are simple and undivided. The species is highly variable; plants can be very substantial with leaves over 30 cm (1 ft) long in a single rosette or may be small multi-rosetted plants with the stature of *M. quintuplinervia*. The flower colour can vary from the most perfect and pure sky-blue to a muddy purple. There are some quite attractive purple-flowered varieties and some with darker blue flowers. The flowering scapes can be 45 cm (18 in) or more long, spindly with a thin-petalled purple flower pendent at the top or they can be thick stemmed with a sky-blue flower 7.5 cm (3 in) or more across. There is a regrettable tendency in this species for the worst flower forms to be on the most polycarpic plants. They usually flower as biennials and most plants are 15–20 cm (6–8 in) across at flowering time and produce only a few flowers and rarely as many as half a dozen.

The seed is larger and more elongate than *M. betonicifolia* and resembles that of *M. grandis* and care is needed with it for two reasons. It sometimes refuses to germinate at all and it is possible that, with some forms, autumn sowing is desirable, as it is with both its close relatives *M. punicea* and *M. quintuplinervia*. It is also poor in percentage germination, even when it does condescend in spring, but this is unpredictable. The answer with a lot of seed of a good form is to sow thinly in a really large pan. If a good germination occurs you can exercise some iron willpower and throw most away or prick them on for a deserving nurseryman. If only a few germinate then you will still have a viable planting for the garden. They should be rapidly pricked on as with other species to a good rich compost into a 10 cm (4 in) pot and then planted on out into the garden. This is the point to beware, they are perfectly reasonable and accommodating at this stage, becoming fully dormant over winter—and very often that is the last you will see of the especially good sky-blue forms (particularly plants known as 'Bailey's Form'). It seems most likely that they object to winter wet and that it is not a spring problem with slugs as it can be with other dormant species. A simple 20 cm (8 in) tent cloche with a handful of dead leaves is all that is required. I do not believe the plant lacks hardiness, but there is no doubt that uncovered there is a real risk of losing the lot—as I know to my own discomfort. This may not be true of all forms but, as they are small plants, fragments of glass held together by cheap simple clips is all that is required. The species is not common in cultivation and one should not take undue risks with good forms.

Some forms are almost invariably monocarpic and die after flowering, very often as biennials. In other forms a proportion will be polycarpic, especially if they are rapidly grown in good substrates. There is a variable amount of seed set and once again the depressing phenomenon is that the most seed is often in the worst forms.

There are two basic forms in cultivation. The monocarpic form with sky-blue flowers is called 'Bailey's Form', which is all the commemoration the poor chap has now that *M. baileyi* is *M. betonicifolia*. The other more perennial form usually has purple flowers. This is in fact a simplification since though 'Bailey's Form' is utterly distinct, the rest are very variable with a proportion being quite attractive robust flowers and occasionally of a very good blue. Donald Lowndes reported such forms in the wild. 'Bailey's Form' is extremely rare and my last planting set almost no viable seed. It is a pity that we could not have combined the polycarpic nature of some forms and produced the ideal of a sky-blue perennial plant. A fertile hybrid of 'Bailey's Form' with *M. sherriffii* (it has a very poor flower) is very polycarpic. This does suggest hybridisation could be valuable in improving the constitution of difficult but desirable plants.

M. simplicifolia requires a rich deep pocket in the rock garden especially for the dwarfer perennial forms and a rich scree in full sun for the monocarpic forms with built in mist (see p. 36). They would grow in the front of an enriched peat bed provided it was well drained in winter and they could be covered. In Scotland they will take full sun, but some shade, especially from the mid-day sun, might be desirable further south.

It is widespread in the wild from central Nepal into Tibet and over a range of altitudes from lower alpine meadows to high screes. It is likely to be very variable, as indeed we know it is, and there is much scope for collecting more material from the wild. Plants have recently been introduced by seed from Sikkim by the Alpine Garden Society expedition as well as by Sinclair and Long from Bhutan. Both these forms are at present widely cultivated and some are reported to be good forms of a good colour.

M. SINUATA

A blue-flowered monocarpic poor relation of *M. latifolia*, not in cultivation.

This is distinguished from *M. latifolia* and *M. aculeata* by long narrow lobed leaves with the usual complement of spiny bristles and a long narrow seed capsule. Prain in his original description recognised two varieties, one of which Taylor subsequently placed in the species *M. horridula*. All this firmly places the plant as being very similar to the three species commonly in cultivation. Taylor describes it as having few flowers of variable colour between blue and violet and that only one flower opens at a time. It was flowered by Harley who was awarded a certificate of merit (a local award of the Scottish Rock Garden Club) but it was only transiently in cultivation.

This species occurs from central Nepal to Bhutan at the altitude of the alpine meadows. It is not clear what particular niche this species is adapted to since *M. horridula* occurs in these same regions with a variety of different forms from the highest altitudes that flowering plants occur. It is possible that when a wider range of material is known that this species will not be maintained with full specific rank.

M. SMITHIANA

A polycarpic yellow-flowered species related to *M. villosa*.

It comes from the Yunnan/Burma border and is very similar in all respects to *M. villosa*. It is distinguished in growth by the tri-foliate leaves as opposed to the lobed leaves of *M. villosa* and in seed by the much rounder seed capsule. Two collections of material from the wild, one by Farrer in Burma, show it to be very similar in growth form to *M. villosa* and it would no doubt be an equally desirable woodland plant with rich yellow flowers. This species is described from stony alpine meadows and possibly might be happier in more open conditions than *M. villosa*. One has to say that, in horticultural terms, since *M. villosa* is so delightful and amenable, that if this species arrives in cultivation it will be largely of interest only to the specialist.

M. SPECIOSA

Yet another *M. horridula* relative with azure-blue flowers and other desirable characteristics that has only fleetingly been in cultivation.

It is the species that is referred to in the title of Kingdon-Ward's book *The Land of the Blue Poppy*. This is a species from the Tibet/Yunnan border and its virtues include the shot silk flowers of azure-blue associated with good forms of *M. aculeata*, a very generously floriferous nature, a nicely dissected foliage with ginger-red hairs and a reputed scent of hyacinth. The latter feature of a scent would be a crowning glory to an already virtuous genus. The species is likely to be mainly biennial and forms a robust flowering stem up to 60 cm (2 ft) in height with a dozen or more flowers out at once. These are large, up to 10 cm (4 in) in diameter, and of a red-purple or at best a brilliant azure-blue with the form and texture of shot silk. Forrest described them as intensely fragrant which is a characteristic that is absent from any *Meconopsis* at present in cultivation. The dissection of the leaves is much greater than in almost all specimens of *M. aculeata* and geographically the two species are widely separated and not likely to be confused in the field for this reason.

The species has been attempted in cultivation a number of times with both Forrest and Rock sending home seed which germinated. There was difficulty in bringing them to flowering which suggests that this species may not be easy. It is described as coming from limited geographical areas on the south-eastern border of Tibet and China and occurring on both acid and alkaline screes up to the highest altitude that vegetation occurs. This narrow range and the high altitude make it similar to the very good forms of *M. horridula* from Sikkim and, interestingly, *M. speciosa* shows the same reduction to flowering from basal scapes that is found in this species. It is reported from dry south-facing crevices. The high altitude form of *M. horridula* is also very difficult to grow and the reasons for this are far from clear. It may be that it is just a question of finding the right combination of dry dormant winter conditions and spring and summer moisture, but it may be that quality and intensity of light are also significant and we have done little in horticultural terms to come to grips with this and other environmental variables associated with high altitude plants. The ability to build small computer-controlled environments is not beyond the wit or pocket of amateur plantsmen and it is from here that the discoveries will be made as to how to grow plants from very specialised environments.

M. SUPERBA (Plate 18)

A truly superb evergreen monocarpic plant with large white flowers, AM 1940.

This species is very similar to *M. regia*, except that it has white flowers and silver leaves, and perhaps has slightly fewer flowers on the same size plant. It is magnificent both in winter leaf rosette and also in summer when it finally

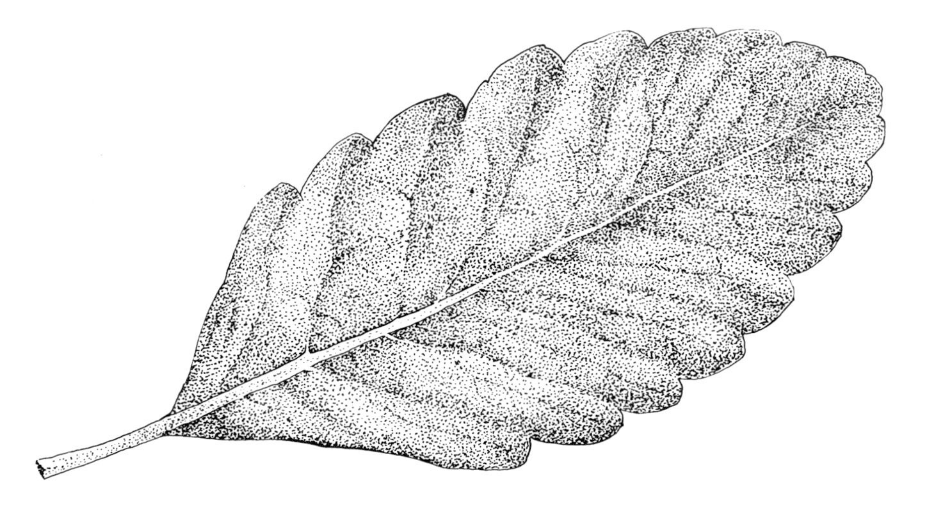

25. *M. superba* leaf from immature plant (the white flower colour and grey leaf foliage separate this from very similar adult leaves of *M. regia*)

flowers. The leaves, like those of *M. regia*, show no lobing or dissection, only the faintest of a saw-toothed edge. The rosette expands gradually over three or four seasons until it is up to 90 cm (3 ft) across, and the silver hairs adpressed to the leaf give it a silky sheen. The flowering spike emerges early (with the possibility of dire consequences) and will eventually reach 2 m (6 ft). The flowers arise on pedicels from the leaf axils of the stem. They tend to be borne singly at the top of the stem but they can be up to five lower down, though three would be the general rule. The flower is large, up to 9 cm (3½ in) across and of the purest white. The stigma is a deep purple-black and a strikingly beautiful feature of the flower. The whole plant is just less substantial in all aspects to *M. regia* and the leaf rosette is sufficiently distinct in smooth outline and colour to be distinguished at any stage beyond the seedling from all other species.

The seed is usually set fairly generously but cold weather in early spring may produce a poor set in the upper flowers and the seed pods are prone to rotting in a wet summer. It is slow germinating, and as there are difficulties in the cultivation of this species, if in doubt, sow thinly in a large pan when plenty of seed is to hand. The seed takes up to two weeks longer than some other species even if grown in warmth; as it is slow growing it is essential to start this species as soon as is possible and by mid-February at the latest. This species is one of the last to be pricked on if all are started together and even then the seedlings may have true leaves less than 2.5 cm (1 in) long. This, with *M. betonicifolia*, is the most prone to damping off at the seedling stage. A very dilute fungicide may help but great care is needed. Growth will remain slow but the plants at this stage at least have the merit of looking nice and neat and not the rather straggly things that many of the other rosette species give rise to at the seedling stage. The plants will in general be half the size of related species by the time of planting out in mid-August. This is perhaps the one species that might be no worse for overwintering in the pot. They are very greedy feeders and they must be planted out into a good rich soil; if the autumn is mild and wet they will be 15 cm (6 in) across by the onset of winter. I think this species must be covered, but again it is essential to use individual tent cloches as they will surely die if they become dry over winter. A large flat pane of glass raised over a group of plants is disastrous. Dry leaves normally blow under the cloche and will give added protection especially from cold desiccating winds. I have never seen winter frost damage with this species but have the suspicion that prolonged temperatures below –10°C (14°F) would not be appreciated without snow cover.

The rosettes will now grow rapidly in the second year and flowering at this stage would be very unusual. You would be very fortunate to bring all of a group of seedlings to flowering and one should probably allow 20–30 per cent more at planting out than you want to flower. They give no problem in summer if in a rich well-fed bed and will take dry or wet weather. The following winter they again require covering and, as they may now be well over 45 cm (18 in) across this does require larger pieces of glass. They again need dead leaves and bracken over them and if really hard weather sets in late in March, once they are stirring from winter resting conditions, conifer branches or bundles of bracken should be added as frost protection. The flower buds are very susceptible to even a light frost. Some plants will flower at three years, the same sort of number at four years and the stragglers at five years. This is quite useful since it gives a continuity of seed over several seasons. A central stem arises from the rosettes and can elongate to 2 m (6 ft) and in mid-May the beautiful chalices of white unfold to reveal the black stigma. This may seem a rather fussy procedure for producing the flowering plants, and covering the rosettes in winter certainly detracts from one of the lovely features. I am sure in many gardens in many years they would survive unprotected and only occasionally would a late frost do serious harm. In a large garden the answer is to protect a few to ensure seed for the future and let the rest take their chances. I have certainly grown them well without cover. Frost damage may not mean loss of seed since the lower flower buds may not have been in advanced enough stage to be damaged but the flowering spike would be utterly spoilt. This species is undoubtedly worth the effort and as the knowledge of how to cultivate it exists, it is worth trying anywhere.

This species was introduced into cultivation in the mid-1920s from an unknown source but only became widespread with a major re-introduction by Ludlow and Sherriff in 1933 from western Bhutan. It occurs in alpine meadows at about 3,900 m (13,000 ft) in western Bhutan and neighbouring Tibet and is a plant from above the snow line. In spring, no doubt once the snow has melted, the wild habitats are fairly frost free and this indicates that this plant is hardy but adapted to winter snow cover in an area where this is deep and predictable.

M. TAYLORII

A newly described relative of *M. regia* that was only fleetingly in cultivation.

This species was collected by the Stainton, Sykes and Williams Expedition to Nepal in 1954 under the numbers SSW 8506, 8507 and 8611 in the Annapurna Himal on open slopes at around 3,900 m (13,000 ft). It was described by Williams in 1972. It is distinguished from *M. regia* by having a seed capsule twice the length at 6 cm (2½ in), pink flowers, longer hairs with filaments (technically barbellate) and a pollen morphology that is radically different to those of *M. regia* and *M. superba* when viewed under the scanning electron microscope. The stigma is a dark brown (as opposed to purple in *M. regia* and *M. superba*). The pollen morphology was part of a study by Sir Douglas Henderson (of the Royal Botanic Garden, Edinburgh) which showed an unexpected diversity between species that are closely related by all other criteria. This is not the place to become involved in taxonomic arguments but it sees to me that if very closely related species that are inter-fertile have radically different pollen then the characteristics of the pollen are a later and secondary development and tell nothing about relatedness. This species was clearly a nice pink *M. regia* and although one can do nothing but applaud the idea of honouring Sir George Taylor, one has to wonder whether this really is a valid species. It has disappeared without trace and presumably interbred with all the other lovely SSW forms before it was even described as a species.

M. TORQUATA

A blue-flowered monocarpic species with deciduous foliage that has only fleetingly been in cultivation.

This species occurs only in Tibet and is a close relative of *M. discigera* and these two species (plus the newly described *M. pinnatifolia*) differ from all others in the genus by the presence of a flat disc surmounting the ovary. The leaves are about 15 cm (6 in) long, narrow and strap-like with a variable amount of lobing at the tip and are thus very similar to that of *M. discigera*. The plant dies back to a resting bud, still visible in winter, lying amidst the dead remains of the previous year's foliage. The plant re-expands in spring and gradually enlarges though it may take a number of years to flower, possibly up to five. The flowering spike reaches 50 cm (20 in) with flowers on very short stems which is attractive. The flowers were originally reported as pale red by Taylor from specimens collected by Walton in 1904 but since that time it has been much better characterised by Ludlow and Sherriff, both of whom were stationed in Lhasa and became very familiar with the plant during that period. The species has been in cultivation briefly and was described as such by Gen. Murray-Lyon who gardened in Perthshire. The flowers tended not to open properly and it did not set seed. It is clearly a desirable plant, as is *M. discigera* and it is probably at least as difficult. The species requires very seasonal dry and wet periods. In the native habitat of the harsh Tibetan plateau it is rarely subject to cold damp weather particularly in autumn and this is likely to be the danger period. Cultivation is likely to be similar to that described for *M. discigera*. The plant is monocarpic and keeping such plants in cultivation on a continual basis is fraught with difficulty where they have to be constantly renewed from seed. It is possible, I suppose, that one day the collection of wild seed from what are effectively weed species in these remote areas may become a local industry and a regular supply of seed may become available.

M. VENUSTA

An attractive relative of *M. horridula* with pale blue or purple flowers all borne on basal scapes. It has never been in cultivation.

This plant is one of a group with *M. horridula*, *M. impedita* and *M. pseudovenusta* that Taylor sorted out by creating the new species of *M. pseudovenusta*. They differ on certain botanical features and it must be at least conceivable that if a really comprehensive range of material was available that the four species could not stand. This species is distinguished by having all basal-scaped flowers with never more than four petals and a very long narrow seed capsule. It is separated from the similar *M. impedita* by its glabrous and glaucous foliage and generally more robust nature. The plant has highly divided leaves forming a single rosette from which eventually up to 15 flowers emerge on basal scapes up to 18 cm (7 in) long. It is clearly delicate and well proportioned and as such is a desirable garden plant. Taylor questioned whether this species was perennial—if it were, it would be unique in the series—basing his suggestion on the evidence that Forrest had collected a plant with flowers and seed pods. Taylor also noted that there was much dead leaf remains on many of the herbarium plants and that this might either indicate a perennial nature or that it took some years to come to flowering. Taylor suggested the seed and flowers on the same specimen were indicative of flowering over two seasons but this need not be the case. A significant proportion of *M. horridula* will, if a damp spell comes after a dry summer, throw a few more flowering basal shoots when seed has already set from the spring flowering. This might certainly appear in a herbarium specimen to represent two years' flowering. On very rare occasions a single plant of *M. horridula* and *M. latifolia* will flower a second year. It is more likely however that this species takes three or more years to reach flowering. If future material was obtained, and it was found to be truly polycarpic, it would be an especially valuable characteristic. It might even have potential for breeding a race of perennial plants based on the best forms of *M. horridula* and *M. latifolia* suitable for drier rock garden situations.

M. VILLOSA

A common and easy perennial wildling with yellow flowers.

This is a charming herbaceous species that is not invasive either in seeding behaviour or in growth. In truth it is more sinned against than sinning since one so often sees it planted towards the back of the woodland glade-type of bed or relegated to the place where not much else will tolerate the overgrowth of a vulgar hybrid rhododendron in the peat bed. The shape of the yellow flowers at once puts one in mind of *M. cambrica* but the plant is distinguished, at least botanically, by the dense hairiness of the base of the plant. The leaves are lobed and rounded on emergence in spring but these rapidly wither as the unbranched flowering stems emerge. These bear at intervals of several centimetres large similar-shaped leaves on long stems. The flowers emerge singly from the axils of the upper leaves on stems up to 15 cm (6 in) long. It is not truly herbaceous since it is at least partially evergreen with what is effectively the new growth emerging in autumn and overwintering. The seed capsules are long and narrow and are distinctive in that they open by slitting sideways up to half the length of the capsule. It is relatively easy from seed but like *M. cambrica* this does not always germinate very predictably; probably sowing a pan soon after harvesting will produce the most reliable germination. They form slightly straggly seedlings and need pricking on young and great care needs taking when potting on because they resent disturbance. If they are planted out in a woodland soil in semi-shaded or in the south, even in a fully shaded position, they will grow on and flower the following year and every year after without any particular problems.

It usually freely sets seed even though only a single plant is present, but perhaps this explains the unreliable germination of the seed. I have rarely seen self-sowing seedlings. A large plant is readily divided and this is probably best done in spring. It will not stand splitting up into many pieces since it is intolerant of too much disturbance. It is not a greedy feeder like many of its perennial and monocarpic relatives but can stay in the same spot for many years. It is probably the one species of meconopsis that will respond to top dressing with leafmould since this is what it is adapted to in the wild. There are a number of forms in cultivation one of which has particularly good yellow flowers and was associated with Sherriff's garden at Ascrievie; indeed, this is a species collected by Ludlow and Sherriff.

It is quite widespread in the Himalayas and was first collected by Hooker approaching a century and a half ago (as *Cathcartia villosa*) in Sikkim but has been collected hundreds of kilometres further east in Bhutan. It occurs over an altitude range of 1,500–2,700 m (5,000 ft–9,000 ft) upwards and this may indicate that some strains are likely to be hardier than others and perhaps there is variability in the degree to which it becomes winter dormant. It is very clearly a woodland plant in the wild and occurring in scarce open glades in quite dense forest.

M. VIOLACEA

A monocarpic evergreen rosette-forming species with blue-purple flowers and distinctive foliage, now rare or extinct in cultivation.

This species, from the number of illustrations that exist of it from the time it was common in cultivation, was clearly an attractively different plant from those that we cultivate now. The rosette of the species is formed of long, fairly narrow leaves that are deeply dissected but each part is rounded (see Figure 14). This produces a most attractive fern-like appearance. It can be distinguished from *M. gracilipes* which is equally dissected by the rounded as opposed to narrow lobes. It differs from *M. napaulensis*, *M. paniculata* and *M. robusta* in that all these species have an irregular amount of dissection and the lobing form is variable. The nearest plant to it is *M. napaulensis* 'Wallich's Form' but in this plant the dissection of the leaf is not so complete nor the lobing so rounded. A flowering spike 1.2–1.5 m (4–5 ft) high arises and is usually shorter than all the other related species. Many of the flowers are borne singly on 7.5 cm (3 in) stems from the upper leaf axils but in cultivation, at least, there may be up to three lower down. The flower is variously described but clearly is another of the lovely shot silk purple-blue mixture. The only other blue-flowered species likely to be confused with it is 'Wallich's Form' of *M. napaulensis*. This latter plant has clear pale blue flowers at its best but can be a dreadful muddy purple that could not possibly be mistaken for the purple-blue described when this plant was common in cultivation. It is likely that this plant flowers at two or three years from seed but now sadly seems to be lost to cultivation. Regrettably it comes from a currently pretty inaccessible part of the Himalayas.

It is certainly not straightforward in cultivation and was not recommended for gardens in the south of Britain even when it was common because of particular requirements of summer humidity. All its close relatives that we have in cultivation will tolerate a great deal of dry weather if kept well-fed and especially so if ground watering can be given. This species appears to require a humid atmosphere in summer or it will not survive. It certainly comes from an area (the eastern end of the Himalayas on the Tibet/Burma border) where it will be subject to the full effects of the monsoon. It is likely to be the one winter rosette-forming species that is fussy as to a strict winter dry, summer wet regime and no doubt was lost when it was given average treatment for its type in drought summers. A misted bed in summer and cover in winter will be ideal if it again becomes available

This species was discovered by Kingdon-Ward from Upper Burma and south-eastern Tibet at altitudes from 3,000–4,000 m (10,000–13,000 ft) and is probably a plant from the edges of woodland and alpine meadows.

M. WUMUNGENSIS

Another newly described Chinese species found on Mt Wumung near Kunming in Yunnan.

It is described as a small blue-purple annual species. It is related to *M. lyrata* which is also insignificant and has been described as an annual. There is no real reason why some should not be annuals but seeing is believing and, without further evidence, I suspect that it is a small biennial. It grows among damp rocks at approaching 4,000 m (13,000 ft). It is a rare plant with incomplete material obtained. It is close to areas currently being botanised so it may possibly turn up. If it is an annual it should not be difficult.

M. ZANGNANENSIS

The fourth new Chinese species which appears to resemble a small *M. simplicifolia.*

One presumes that it also closely resembles *M. quintuplinervia* and as it is described as a perennial with sky-blue flowers it is clearly a gem. The species comes from high alpine pastures above 4,000 m (13,000 ft) on the Tibetan border east of Bhutan.

6 *Hybrids*

The creation of hybrids can be a slightly emotive subject when such a classic genus is involved. The form of many species of meconopsis and the size of flowers make a breathtakingly beautiful plant. Flowers on *M. integrifolia* have been described as 28 cm (11 in) across, and a good form of *M. napaulensis* may have up to 300 or 400 flowers on it over six weeks. A great deal of improvement could be made with them as garden flowers if rigorous selection of good forms and colours were made. Most of the genus are fairly self-sterile and hand pollination needs carefully and repeatedly doing to ensure a good set of seed between two selected plants. Patience will be rewarded however since, although there is variation in the most carefully pollinated batches of seed, colour and form are in general maintained. Simply weeding out poor plants from the garden the moment they have flowered would do much to improve strains.

It might well be asked what value is there in hybridisation, especially in monocarpic species which may well leave sterile progeny. The answer in some cases must be nothing at all; the cross between *M. napaulensis* and *M. dhwojii* is a sterile shadow of either parent even though it was graced with the name *M.* × *ramsdeniorum*. The hybrid between *M. betonicifolia* and *M. grandis*, named *M.* × *sheldonii*, is however as stunning as it is famous.

There is great potential in hybrids to do two things. First, there is scope to produce a race of variably coloured dwarf perennials based on *M. quintuplinervia*, and second, crosses between the monocarpic rosette-forming species produce perennial evergreen plants which can have very tough constitutions.

The *M. quintuplinervia* crosses are best exemplified by that with *M. integrifolia*. This has been made a number of times, first as *M.* × *finlayorum* by the Knox-Finlays at Keillour which was validly described by Taylor. Mike and Polly Stone of Fort Augustus have re-made the cross and called it *M.* 'Askival Ivory' and I have made it myself. The progeny have the usual cream-coloured flowers of a blue/yellow cross, are of a very good size and the plants do not exceed 30 cm (1 ft) even at flowering. A proportion of the plants are highly perennial though the majority may be monocarpic and short-lived. Selection and vegetative propagation of good clones of this cross is at present being undertaken and will be an important addition to easy rock garden meconopsis when they are available from nurseries. There is potential to use *M. sherriffii*, *M. punicea* and *M. simplicifolia* with *M. integrifolia* and produce strains of variously coloured dwarf perennial hybrids. In some cases these have already been attempted and sometimes a cross between two doubtfully polycarpic parents produces offspring that are very perennial.

The second type of cross is just as exciting in potential. The theory of it is difficult since the chromosome numbers of the parents do not fit in the least. The cross is made between a monocarpic rosette-forming species such as *M. napaulensis*, and a perennial species such as *M. betonicifolia*. This cross has been made half a dozen times between a number of different species. I have experience of a cross I made between *M. napaulensis* (a yellow form, presumably hybrid itself with *M. regia* or *M. paniculata*) and a form of *M. grandis*, with the latter as the pollen parents. The seedlings initially resembled *M. napaulensis* but by the second summer they all became multi-rosetted and every one of them is totally perennial and has the toughest constitution of any meconopsis I have ever grown. They can be divided into a dozen pieces every year; ten years ago when I produced this I gave many away, and I still find them growing happily, often in the most neglected and starved conditions. None of the offspring was particularly floriferous, making vegetative increase instead and the colour was the expected yellow-cream. The fact that they are remarkably perennial, will stand dry and starved conditions and are attractively evergreen with total frost hardiness shows the very considerable potential from these crosses.

The list which follows indicates what has been made; where the hybrids have not been named they will be found under the relative parents.

M. 'Askival Ivory'

A deliberate cross between *M. integrifolia* and *M. quintuplinervia*, made by Mike and Polly Stone of Fort Augustus. The best forms are dwarf with cream flowers and are a very desirable addition to the rock garden since at least some are perennial. This is the same hybrid as *M. ×finlayorum*.

M. × auriculata

This is an obscure hybrid that occurred in a batch of *M. aculeata* seedlings, although clearly they were rogue and not related to that species. Taylor decided that the evergreen nature of the plant, its pale yellow flowers and its overall resemblance to *M. betonicifolia* made it a likely hybrid between that species and *M. paniculata*. The description of the plant does make this likely, though it is possible that another yellow-flowered rosette species was involved. The perennial evergreen nature of this type of hybrid makes it potentially valuable though there is no recent record of this plant.

M. × beamishii

A cross probably produced many times between *M. integrifolia* and *M. grandis*. It first flowered in 1906 and was raised by R.H. Beamish near Cork. The flowers are cream but some are reported to have purple streaks at the centre of the flower. The hybrid looks like *M. grandis* and as such can always be distinguished from the similar cross *M.* × *sarsonsii* which has *M. betonicifolia* as the blue parent. The hybrid is fertile and most of the progeny are monocarpic but a tiny percentage are polycarpic. These last are usually of a better form since so many of the monocarpic flowering spikes are top heavy and the whole plant ungainly. Different batches of seed may well produce different proportions of polycarpic plants, and effort is worthwhile to select for this, since the very large cream flowers are a good addition to a meconopsis collection and such plants can then be propagated vegetatively.

M. cambrica × *M. quintuplinervia*

This hybrid was reported 60 years or more ago and would be a cross between species with dissimilar chromosome counts. It has not been recorded since, nor has any other with the native British wildling, but nothing seems impossible in this genus.

M. × *cookei*

A hybrid first raised by Andrew Harley of Devonhall between *M. punicea* and *M. quintuplinervia*. It is a muddy compromise between the two. It is perennial with muddy purple-red flowers drooping on tall basal scapes. It has almost died out in cultivation in the last few years and very few plants still exist.

M. × *coxiana*

This cross, commemorating the late E.M.H. Cox who accompanied Farrer on his last trip, is a hybrid between the rosette-forming species *M. violacea* (now no longer in cultivation) and *M. betonicifolia* (pollen parent). An evergreen perennial but not now in cultivation.

M. 'Crewdson Hybrids'

see *M.* × *sheldonii*.

M. × *decora*

This is a white- or blue-flowered plant raised by T. Hay, ostensibly from wild Tibet seed. It seemed likely to Prain and subsequently to Taylor that it was a hybrid between *M. napaulensis* and *M. latifolia* (and thus could not have come from Tibet). It occurred spontanteously in a number of places in the early 1920s. This hybrid was interesting in that it flowered without setting seed for a number of years and the presumed parents are both monocarpic. There is, however, need to confirm this unlikely cross.

M. × *finlayorum* (Plate 19)

Cross between *M. integrifolia* and *M. quintuplinervia* produced by Knox-Finlays at Keillour. See *M. 'Askival Ivory'*.

M. grandis × *M. latifolia*
This hybrid was reported by Farrer in *The English Rock Garden* as a perennial plant that occurred in a garden in Ireland. If this evidence is accepted then it further confirms that nothing is impossible in this genus, even if such hybrids are very rare indeed.

M. grandis × *M. simplicifolia*
This was a hybrid raised by F.C. Puddle at Bodnant in Wales. A proportion of the progeny were polycarpic and had flowers on basal scapes up to 1.2 m (4 ft) high. It clearly resembled the hybrid *M.* 'Houndwood' and it is possible that *M. simplicifolia* is involved in other garden hybrids and maybe even in good wild-collected forms of *M. grandis* and *M. betonicifolia.*

M. grandis × *M. sinuata*
This is the same as the hybrid with *M. latifolia.*

M. × *harleyana*
This is the 'Ivory Poppy' and was also described as *M. simplicifolia* var. *eburnea.* It is the cross between *M. simplicifolia* and *M. integrifolia* and has large cream flowers arising on a basal scape. It occurred spontaneously in the garden of A. Harley of Devonhall in mid-Scotland. It was polycarpic and appears to have been fertile, and it also occurred in the wild. It is no longer in cultivation but the huge flowers of some forms of *M. integrifolia* show the great potential of this species for hybrid work.

M. 'Houndwood'
This cross was introduced by Alec Duguid, of Edrom Nurseries just east of Edinburgh, and was found as a natural seedling in the garden of a hotel of that name. It closely resembles a robust form of *M. simplicifolia* and is a most excellent hybrid. The parents were emphatically reported to be *M. betonicifolia* and *M. quintuplinervia* as no *M. simplicifolia* grew anywhere near, and this must be accepted. Margaret and Henry Taylor, who garden at Invergowrie near Dundee, have re-made this cross. Their un-named hybrid, though very nice, is quite unlike *M.* 'Houndwood', and resembles *M. quintuplinervia*, so the mystery must remain. The Taylors have also collected and grown on viable seed from *M.* 'Houndwood' although to date my own plants have been sterile.

M. × *hybrida*
This is the cross between *M. simplicifolia* and *M. grandis* made by Puddle.

M. 'Kingsbarns Hybrids'
A single plant of *M.* × *sheldonii* produced viable seed in quantity. The resulting progeny, which are also highly fertile, are very variable in form and

colour and clearly are the result of a back cross which could have been either *M. betonicifolia* or *M. grandis* since both were nearby. Seed of *M.* × *sheldonii* is often offered but never true, though seed has occasionally been reported in the past. I think this hybrid, which is very robust, may be valuable in parts of the world where vegetatively propagated material of *M.* × *sheldonii* itself is unobtainable. I have re-named them (Cobb 1989) since *M.* × *sheldonii* is only applicable to the original hybrid and not back crossed progeny.

M. 'Miss Dickson'

This is sometimes simply described as a white form of *M. grandis*. Such plants do exist, though they are very rare, and seed often offered as *M. grandis* var. *alba* is invariably *M. betonicifolia* var. *alba*. I have seen this form at Keillour, and a more robust form at the late Bobby Masterson's garden at Cluny in Perthshire and once an *M. grandis* seedling of my own was white. For all this, *M.* 'Miss Dickson' was described as a hybrid and presumably of the same cross as *M.* × *sheldonii*. It docs not look likc *M.* × *sheldonii* (assuming the Keillour plant was the real *M.* 'Miss Dickson') and closely resembles *M. grandis* itself. A cross between *M. grandis* and the white *M. betonicifolia* might produce a proportion of white *M.* × *sheldonii* but so far my attempts at this cross have failed.

M. × *musgravei*

This is an interesting cross between *M. betonicifolia* and *M. superba* produced by C.T. Musgrave and described in 1933. The hybrid was white-flowered and polycarpic and one presumes was also evergreen. Taylor described it as fertile which is extraordinary since the chromosome counts of the two parents are quite different and the species not closely related. He later reported it to be a difficult plant and it no longer exists.

M. napaulensis × *M. grandis*

This is my own un-named cross (1984). The plants are all very similar. They are evergreen, immensely robust but poor in producing flowering spikes. The hybrid shows the immense potential of this type of cross.

M. 'Ormswell'

See *M.* × *sheldonii*.

M. × *ramsdeniorum*

A cross between *M. dhwojii* and *M. napaulensis* with yellow or pink flowers. It was produced nearly 50 years ago by Sir John Ramsden. It is sterile and less good than either parent and occurs spontaneously in mixed plantings of the parents, which is clearly undesirable.

M. ' × regia'

It seems very unlikely that true *M. regia* is still in cultivation. It tends to hybridise with both *M. paniculata* and *M. napaulensis*. These plants often closely resemble *M. regia* in being very robust with a smooth outline to the leaves but lack the deep purple-black stigma.

M. × *sarsonsii*

This is a well-known hybrid between *M. betonicifolia* and *M. integrifolia* first produced by Sarsons at East Grinstead in Sussex. Nearly all the seed that circulates at present of this fertile cross is not correct and will be found to be the very similar *M.* × *beamishii*. There are few polycarpic forms of this hybrid in cultivation and they are very desirable. They can be distinguished from *M. beamishii* by exactly the same characters as *M. betonicifolia* can be distinguished from *M. grandis* (see Figure 9). The hybrid can be polycarpic or monocarpic, but mostly the latter. It is fertile but vegetative propagation is quite possible in polycarpic forms using the same procedures as one would use on *M. betonicifolia* itself.

M. × *sheldonii* (Plate 20)

This is the magnificent cross between *M. betonicifolia* and *M. grandis*. It was first raised by W.G. Sheldon in 1934 at Oxted in Surrey and described by Taylor. The cross has certainly been made a number of times and there are

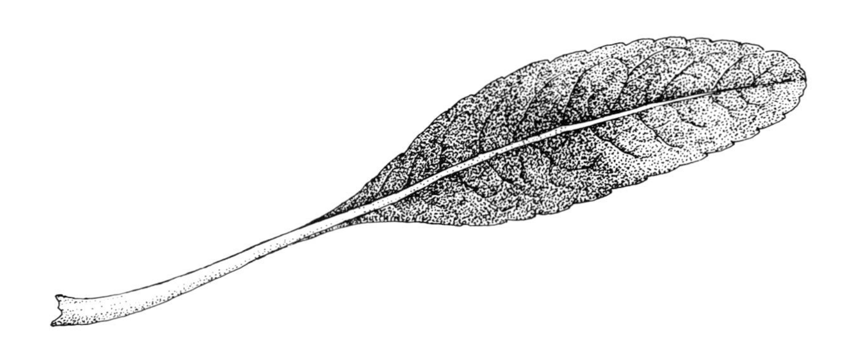

26. *M.* × *sheldonii* leaf

various forms of it in cultivation. Alec Curle of Edinburgh also made this cross and plants given to Edrom Nurseries were named *M.* 'Ormswell'. Other plants from this same cross were passed via Hugo Patten and Marjorie Dickie to the Slieve Donard Nursery in Ireland who called it *M. grandis* 'Prain's Variety'. This invalid name was later changed to *M.* 'Slieve Donard'. There is a pale-flowered form of *M.* × *sheldonii* called *M.* 'Aberchalder Form'.

M. 'Crewdson Hybrids' are presumably the same cross but perhaps with a different form of *M. grandis* (remember that the Nepal and Sikkim forms of *M. grandis* are quite distinct). Cicely Crewdson was a famous grower after the Second World War. Crewdson hybrids have smaller flowers and tend to be a darker blue. The original *M.* × *sheldonii* is still in cultivation (I have been given plants of it).

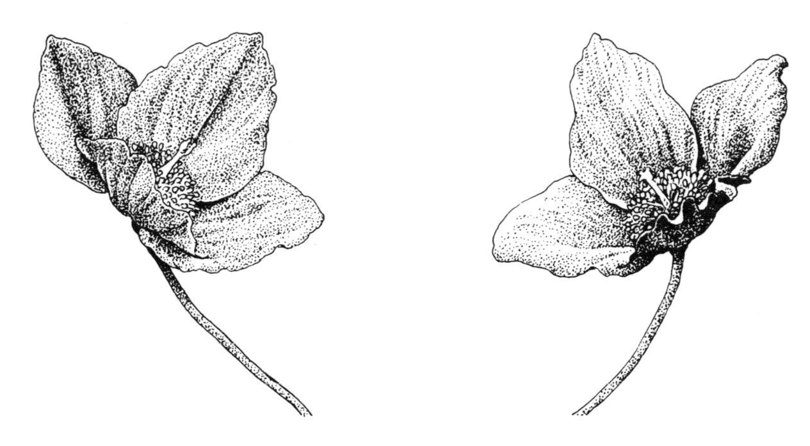

27. Comparison of *M.* × *sheldonii* (left) and *M.* 'Crewdson Hybrids' (right); the latter is slightly more delicate with less overlapping petals

M. sherriffii × *M. simplicifolia*

This hybrid occurred spontaneously between *M. sherriffii* and a plant of the monocarpic 'Bailey's Form' of *M. simplicifolia.* The flower is very small and a disgusting muddy shade. The interesting feature of the plant is that both parents are difficult and only *M. sherriffii* shows any tendency to be polycarpic, but the hybrid is most perennial and tough. It shows the potential for introducing perennial nature and robustness. This hybrid is fertile and thus has the potential to produce back crosses.

M. sherriffii × *M. superba*

This a cross recently made by Margaret and Henry Taylor of Invergowrie. Eventually these crosses will produce desirable perennial evergreen plants. This cross has yet to flower.

M. simplicifolia × *M. paniculata*

Taylor reported this as having occurred in the last century at Edinburgh. The plant was not described.

M. 'Slieve Donard'

See *M.* × *sheldonii.*

Appendix I
Synonyms and non-valid names

Largely based on the work of Taylor (1935)

'Aberuchill'—a variety of *M. integrifolia* which is not valid.
aculeata—vars. *acutiloba, nana, normalis, typica*. Non-valid varieties of *M. aculeata*.
baileyi—now considered the same as *M. betonicifolia*.
betonicifolia—vars. *baileyi, franchetii*. Non-valid varieties of *M. betonicifolia*.
betonicifolia—var. *pratensis*. Form introduced by Kingdon-Ward. Larger and distinct but probably not reasonably separable from the type.
calciphila—now included in *M. horridula*.
cambrica—vars. *arvernensis, borealis, decipiens, gallica, hispanica, incisa, inornata, nubilosa, oblita, oxyphylla, pyrenaica, tanacetoides*. Non-valid varieties of *M. cambrica*.
cambrica—var. *aurantiaca*, the orange variety.
cambrica—vars. *plena* and *flora plena*, double forms of the orange or yellow colour.
cambrica—'Frances Perry', deep orange variety (not scarlet).
cawdoriana—now included in *M. speciosa*.
compta—now included in *M. lyrata*.
concinna—now considered a variety of *M. lancifolia*; briefly in cultivation.
crassifolia—removed from *Meconopsis*, now *Stylomecon heterophylla*.
diphylla—removed from *Meconopsis*, now *Stylophorum diphyllum*.
duriuscula—now included in *M. horridula*.
eucharis—now included in *M. lancifolia*.
eximia—now included in *M. lancifolia*.
florindae—now considered a yellow-flowered form of *M. lyrata* by Taylor.
georgei—now considered a yellow-flowered form of *M. lancifolia* by Taylor.
glabra—removed from *Meconopsis*, now *Stylomecon heterophylla*.
gulielmi-waldemarii—now included in *M. aculeata*.
henrici—vars. *genuina, psilonomma*. Valid varieties of *M. henrici*.
heterophylla—removed from *Meconopsis*, now *Stylomecon heterophylla*.
himalayensis—now included in *M. paniculata*.
horridula—vars. *abnormis, racemosa, rudis, typica*. Non-valid varieties (not considered consistently distinct by Taylor).
impedita—vars. *morsheadii, rubra*. Non-valid varieties of *M. impedita*.
integrifolia—vars. *brevistyla, microstigma, souliei*. Non-valid varieties of *M. integrifolia*.
lancifolia—var. *limprichtii*. Non-valid variety of *M. lancifolia*.
lancifolia—vars. *concinna, solitariiflora*. Valid varieties.
leonticifolia—now included in *M. venusta*.
lepida—now included in *M. lancifolia* var. *solitariiflora*.
morsheadii—now included in *M. impedita*.
napaulensis—vars. *elata, fusco-purpurea, purpurea, rubro-fusca, typica*. Non-valid vars. of *M. napaulensis*.
ouvrardiana—now included in *M. speciosa*.
paniculata—vars. *elata, typica*. Non-valid varieties of *M. paniculata*.
petiolata—removed from *Meconopsis*, now *Stylophorum diphyllum*.
polygonoides—now included in *M. lyrata*.
prainiana—now included in *M. horridula*.
prattii—now included in *M. horridula*.
principis—now included in *M. henricii genuina*.
pseudointegrifolia—now included in *M. integrifolia*.

psilonomma—valid variety of *M. henricii* but not now considered a species.
racemosa—now included in *M. horridula*; has the distinctive feature of yellow anthers.
rigidiuscula—now included in *M. horridula*.
rubra—now included in *M. impedita*.
rudis—now included in *M. horridula*; form with deep purple spots on leaves, not a valid variety.
simplicicaulis—now included in *M. simplicifolia*.
simplicifolia—var. *baileyi*. Non-valid variety; but currently the good blue moncarpic form of this species is known as 'Bailey's Form'.
simplicifolia—var. *eburnea*. Now known to be a hybrid of *M. simplicifolia* and *M. integrifolia* and called × 'harleyana' from the wild and cultivation. The 'Ivory Poppy'.
sinuata—vars. *latifolia*, *prattii*, *typica*. Non-valid forms of *M. sinuata*.
soulei—now included in *M. integrifolia*.
uniflora—now included in *M. simplicifolia*.
wallichii—non-valid species now included in *M. napaulensis* but also not considered by Taylor to be worth varietal status. Blue or white or muddy purple form, which certainly appears worthy of full varietal status if not re-instatement to full species.
wardii—now included in *M. henricii genuina* (also used in the past for *M. lancifolia*).
wollastonii—now included in *M. paniculata*.

Appendix II
Summary of Taylor's Classification with Additions and Corrections

Sub-genus *Eumeconopsis*	*Cambricae* *M. cambrica* *Eucathcartia*(2 series) (1) *M. chelonidifolia*, *M. oliverana.* (2) *M. villosa*, *M. smithiana.* *Polychaetia*, subsection *Eupolychaetia* (2 series)
Series *Superba*	(1) *M. superba*, *M. regia*, *M. taylorii.*
Series *Robustae*	(2) *M. robusta*, *M. dhwojii*, *M. gracilipes*, *M. paniculata*, *M. longipetiolata*, *M. violacea*, *M. napaulensis*, subsection *Cumminsia* (6 series)
Series *Simplicifoliae*	(1) *M. simplicifolia*, *M. quintuplinervia*, *M. punicea*, *M. zangnanensis.* (Also *M. barbiseta* may be belong to this series.)
Series *Grandes*	(2) *M. integrifolia*, *M. betonicifolia*, *M. grandis*, *M. sherriffii.*
Series *Primulinae*	(3) *M. lyrata*, *M. primulina*, *M. wumungensis.*
Series *Delavayanae*	(4) *M. delavayi.*
Series *Aculeatae*	(5) *M. henrici*, *M. forrestii*, *M. impedita*, *M. venusta*, *M. pseudovenusta*, *M. lancifolia*, *M. horridula*, *M. latifolia*, *M. speciosa*, *M. aculeata*, *M. neglecta*, *M. sinuata*, *M. argemonantha.*
Series *Bellae*	(6) *M. bella.*
Sub-genus *Discogyne*	*M. discigera*, *M. torquata*, *M. pinnatifolia.*

Appendix III
Chromosome Counts

Based on Ratter (1968)

2n = diploid number.

28 *M. cambrica*, *M. chelonidifolia* (Ratter reports 2n = 22 from one source for *M. cambrica*).

32 *M. villosa*.

56 *M. regia*, *M. dhwojii*, *M. gracilipes*, *M. paniculata*, *M. longipetiolata*, *M. napaulensis*; also *M. aculeata*, *M. horridula* and *M. latifolia*.

74 *M. integrifolia*.

82 some *M. simplicifolia* and some *M. betonicifolia*.

84 some *M. simplicifolia*, *M. quintuplinervia*, *M.* × *cookei*.

Some *betonicifolia* were c. 120 and *M. grandis* c. 118. Clearly more work needs to be done on *M. grandis*, *M. betonicifolia* and *M. simplicifolia*, preferably with fresh wild material. More *M. integrifolia* need checking to see if this anomalous chromosome number is consistent.

Appendix IV
RHS Colour Classification of Some Meconopsis

Comparisons were made with the Royal Horticultural Society's colour chart for a number of significant species and hybrids.

M. × *beamishi*—160 C.
M. betonicifolia—116 A or B.
M. cambrica'Frances Perry'—33 A.
M. grandis—83 A.
M. horridula, dark blue form—95 A;
silver form—112 C.
M. integrifolia—12 A.
M. latifolia—111 D or 112 A.
M. napaulensis, pink form—65 A.
M. punicea—45 A or 46 A.
M. quintuplinervia—85 A or 94 A.
M. × *sheldonii*—approximately 118 A.

Appendix V
Fungal Diseases Associated with Meconopsis

The following fungal diseases have been identified infecting *Meconopsis* sp. in cultivation. *Mucilago spongiosa* (smother), *Peronospora arborescens* (mildew), *Phytophthora canetorum* (root rot), *Phytophthora parasitica* (damping off), *Phytophthora verrucosa* (root rot) and *Sclerotinia sclerotiorum* (stem rot and wilt). Meconopsis are also prone to crown rot, where an otherwise healthy plant has rotted away from the roots. This is most likely a bacterial infection derived from adverse physiological conditions of plant growth. It may possibly be a secondary infection caused subsequent to an attack by one of the above fungal pathogens. The fungal infections once well established are difficult to control and are better prevented by good growing conditions and the regular spraying of plants most prone to attack with fortnightly doses of systemic and non-systemic fungicides. Care is essential to make these formulations up to the recommended strengths and no stronger. Crown rot is most likely to infect plants in damp stagnant conditions in autumn or alternating wet and cold spells over winter. Close attention to the surroundings of the plant to alleviate these stresses is the only course and is usually effective if done imaginatively.

Appendix VI
Sources of Seed and Plants

Seed

Scottish Rock Garden Club seed exchange. Secretary, Mrs E. Stevens, The Linns, Sherriffmuir, Dunblane, Perthshire FK15 OLP

Alpine Garden Society seed exchange. Secretary, Lye End Link, St Johns, Woking, Surrey GU21 1SW

The American Rock Garden Society seed exchange. Contact: Buffy Parker, 15 Fairmead Road, Darien, CT. 06820, USA

Thompson and Morgan, London Road, Ipswich, Suffolk IP2 OBA

Chiltern Seeds, Bortree Stile, Ulverston, Cumbria LA12 7PB

Plants

Edrom Nurseries, Coldingham, Eyemouth, Berwickshire TD14 5TZ

Holden Clough Nursery, Holden, Bolton-by-Bowland, Clitheroe, Lancashire BB7 4PF

Jack Drake, Inshriach Alpine Plant Nursery, Aviemore, Inverness-shire PH22 1QS

Select Bibliography

Cobb, J.L.S. (1989) 'Kingsbarns Hybrids', *J. Scottish Rock Garden Club* 21;161

Cox, E.H.M. (1986) *Plant Hunting in China* (Reprint, Oxford University Press)

Fletcher, H.R. (1975) *A Quest of Flowers* (Edinburgh University Press, Edinburgh)

Haw, S.G. (1980) 'Meconopsis in Western China', *Quart. Bull. Alpine Garden Soc.* 48;236

Jingwei, Zhang (ed.) (1982) *The Alpine Plants of China* (Gordon and Breach, New York)

Polunin, O. and Stainton, A. (1984) *Flowers of the Himalaya* (Oxford University Press)

Ratter, J.A. (1968) 'Cytological studies in *Meconopsis*', *Notes from the R.B.G.* 28;191

Taylor, G. (1934) *An account of the genus Meconopsis* (New Flora and Silva Ltd., London)

Thompson, D. (1968) 'Germination responses of *Meconopsis*', *J. Roy. Hort. Soc.* 93;336

Glossary

ADPRESSED Pressed flat to the surface.
ANTHER The part of the stamen that contains the pollen grains.
BARBELLATE Spike-like hairs on the leaves.
BASAL Leaves which arise from the base of the stem.
BIENNIAL A plant requiring two years from seedling to flowering, after which it dies.
CAULINE Leaves wrapping around the stem.
CYME An inflorescence formed of axillary branches terminating in a flower, the central flowers maturing first.
DECIDUOUS Leaves falling off and the plant dying back to ground level, usually in the autumn (used as opposed to evergreen).
ELLIPTICAL Oval in outline.
EVERGREEN Foliage maintained throughout the winter above ground.
FILAMENT The slender stalk of the stamen which bears the anthers.
HYBRID SWARMS Groups of plants of varying shape and colour derived from a number of different species.
LEAF AXIL The point at which the leaf-stalk joins the stem.
LOBED Rounded divisions at the edge of a leaf but not completely dividing it.
MONOCARPIC Plants which take a number of years to reach flowering, but die afterwards.
OVARY The part of the flower containing the ovules and later the seeds.
PANICLE A branched flowering stem.
PEDICEL The stalk of a single flower.
PERENNIAL Theoretically living indefinitely, not dying after flowering, polycarpic.
PETIOLE Leaf stalk.
PINNATE The regular arrangement of leaflets in two rows either side of the leaf stalk or petiole (also called a rachis).
POLYCARPIC Living for many seasons, not dying after flowering, perennial.
RETICULATION A net-like marking on the outside (of seed).
ROSETTE-FORMING An arrangement of leaves radiating from the centre and spreading flat over the ground.
RUFOUS Reddish-brown.
SCAPE Flowering stem without leaves.
SCORCH Burnt dry-looking patches on leaves.
SEED CAPSULE The enlarged and dried ovary now containing the seeds.
STIGMA The front part of the female organ that receives the pollen.
STIGMATIC DISC A wide fleshy part of the receptacle which surmounts the ovary (see illustration page 2).
STYLE The more or less elongated part of the female organ which bears the stigma.
STYLAR DISC Effectively the same as the stigmatic disc.
SUBSTELLATE PUBESCENCE Much branched short hairs.
TRI-FOLIATE Having three leaflets to each leaf.

Index